AF581699

Advanced Optimization Methods and Big Data Applications in Energy Demand Forecast

Advanced Optimization Methods and Big Data Applications in Energy Demand Forecast

Editors

Federico Divina
Francisco A. Gómez Vela
Miguel García-Torres

MDPI • Basel • Beijing • Wuhan • Barcelona • Belgrade • Manchester • Tokyo • Cluj • Tianjin

Editors
Federico Divina
Division of Computer Science,
Universidad Pablo de Olavide
Spain

Francisco A. Gómez Vela
School of Engineerings,
Universidad Pablo de Olavide
Spain

Miguel García-Torres
Division of Computer Science,
Universidad Pablo de Olavide
Spain

Editorial Office
MDPI
St. Alban-Anlage 66
4052 Basel, Switzerland

This is a reprint of articles from the Special Issue published online in the open access journal *Applied Sciences* (ISSN 2076-3417) (available at: https://www.mdpi.com/journal/applsci/special_issues/optimization_big_data_energy_forecast).

For citation purposes, cite each article independently as indicated on the article page online and as indicated below:

LastName, A.A.; LastName, B.B.; LastName, C.C. Article Title. *Journal Name* **Year**, *Volume Number*, Page Range.

ISBN 978-3-0365-0862-7 (Hbk)
ISBN 978-3-0365-0863-4 (PDF)

Contents

About the Editors

Federico Divina obtained his Ph.D. in Artificial Intelligence from the Vrije Universiteit of Amsterdam, and, after that, worked as a postdoc at the University of Tilburg, within the European project NEWTIES. In 2006, he moved to the Pablo de Olavide University. He has been working on knowledge extraction since his Ph.D. thesis at the Vrije Universiteit of Amsterdam. His main research interests focus on machine learning and, in particular, on techniques based on soft computing, bioinformatics and big data.

Francisco A. Gómez Vela received his Ph.D. in Computer Science from the Pablo de Olavide University of Seville, in addition to Computer Science Engineering from the University of Seville. His lines of research are focused on the treatment of information using intelligent techniques, applying machine learning and data mining techniques. He has mainly focused on the analysis of genetic and biomedical data in his research. His research is mainly based on the inference of biological models based on gene networks. In addition, he has recently focused on the research of new big data techniques for the exploitation of different types of data.

Miguel García-Torres is an Assistant Professor at the Escuela Politécnica Superior of the Pablo de Olavide University. He received his BSc degree in physics and Ph.D. degree in computer science from the Universidad de La Laguna, Tenerife, Spain, in 2001 and 2007, respectively. After obtaining the doctorate, he held a postdoc position in the Laboratory for Space Astrophysics and Theoretical Physics at the National Institute of Aerospace Technology (INTA). There, he joined the Gaia mission from the European Space Agency (ESA) and started to participate in the Gaia Data Processing and Analysis Consortium (DPAC) as a member of "Astrophysical Parameters", Coordination Unit (CU8). He has been involved in the "Object Clustering Analysis" (OCA) Development Unit since then. His research areas of interest include machine learning, metaheuristics, big data, time series forecasting, bioinformatics and astrostatistics.

MDPI

Editorial

Advanced Optimization Methods and Big Data Applications in Energy Demand Forecast

Federico Divina *, Francisco Gómez-Vela * and Miguel García-Torres *

Computer Science Division, Universidad Pablo de Olavide, ES-41013 Seville, Spain
* Correspondence: fdivina@upo.es (F.D.); fgomez@upo.es (F.G.-V.); mgarciat@upo.es (M.G.-T.)

Citation: Divina, F.; Gómez-Vela, F.; García-Torres, M. Advanced Optimization Methods and Big Data Applications in Energy Demand Forecast. *Appl. Sci.* **2021**, *11*, 1261. https://doi.org/10.3390/app11031261

Academic Editor: Frede Blaabjerg
Received: 22 January 2021
Accepted: 25 January 2021
Published: 30 January 2021

The use of data collectors in energy systems is growing more and more. For example, smart sensors are now widely used in energy production and energy consumption systems. This implies that a huge amount of data are generated, and need to be analyzed in order to extract useful insights from it. Such Big Data gives rise to a number of opportunities and challenges for informed decision-making.

In recent years, researchers have been very actively working in order to come up with effective and powerful techniques in order to deal with the huge amount of data available. Such approaches can be used in the context of energy production and consumption, considering the amount of data produced by all samples and measurements, as well as including many additional features. With them, automated machine learning methods for extracting relevant patterns, high-performance computing, or data visualization are being successfully applied to energy demand forecasting.

In light of the above, this special issue was proposed in order to collect latest research on relevant topics, and in particular in energy demand forecast, and the use of advanced optimization methods and Big Data techniques. Here, by energy, we mean any kind of energy, e.g., electrical, solar, microwave, wind.

In response to the Call for Papers, eleven articles were submitted to this special issue, and five were accepted for publication. If we look at the techniques used in the accepted articles, we can notice that, in two articles, deep learning techniques were used in order to forecast energy demands. In particular, in [1], a Temporal Convolutional Network architecture was studied on two different pieces of data from Spain: the national electric demand and the power demand at charging stations for electric vehicles. In order to test the proposal, an extensive experimental study was conducted. The proposed model was compared with state-of-the-art deep learning-based techniques with different architectures and parametrization. Results show that the proposed model is competitive and outperformed the models used in the comparisons.

Deep Learning was also used in [2]. In this work, a neuro-evolution approach was used. In fact, the configurations parameter of the Neural Network used were set by using an Evolutionary Algorithm. The resulting architecture was applied to energy demand data registered in Spain over a period of more than nine years, and when compared to state-of-the-art approaches, proves to achieve better results.

In [3], the household sector is considered. In particular, data collected from residential homes of England and Wales are used. Such data are first clustered into three groups, according to the houses' consumption profile, and then a decision tree algorithms is used in order to extract insight regarding equipment that could affect energy efficiency. Authors identified various factors that can be addressed in order to improve energy efficiency in houses.

In the last few years, we have also seen a transaction toward green alternatives for energy generations. In order to face this transactions, the photovoltaic power plants' forecasting problem was tackled in [4]. Authors of this article took into consideration various external factors, like meteorological situations in order to produce the predictions. Authors have focused on data filtration, specifically for the data coming from open sources, like data regarding meteorological conditions. This process have proven to help in obtaining more accurate results.

In [5], the authors analyzed and discussed the parameters space of the multiple seasonal Holt–Winters models applied to electricity demand in Spain. This work studied the stability of the smoothing parameters in the multiple seasonal Holt–Winters models to provide accurate and trusted forecasts. In addition, the authors analyzed the variation of the parameters through different seasonal and trend methods. They argue that double seasonal models provide better predictions than triple seasonal ones. Furthermore, the authors established that, for the accuracy of the forecasts, no more than 5000 observations are required. The results of the performed analysis were limited to the Spanish electricity demand. However, the authors expect similar results when faced with load forecasting in other countries or systems.

Acknowledgments: We would like to thank all the authors and peer reviewers for their valuable contributions to this special issue. This issue would not be possible without their valuable and professional work. In addition, we would also like to take the opportunity to show our gratitude to the Genes editorial team for their work, for which this special issue has been a success.

References

1. Lara-Benítez, P.; Carranza-García, M.; Luna-Romera, J.M.; Riquelme, J.C. Temporal Convolutional Networks Applied to Energy-Related Time Series Forecasting. *Appl. Sci.* **2020**, *10*, 2322. [CrossRef]
2. Divina, F.; Torres Maldonado, J.F.; García-Torres, M.; Martínez-Álvarez, F.; Troncoso, A. Hybridizing Deep Learning and Neuroevolution: Application to the Spanish Short-Term Electric Energy Consumption Forecasting. *Appl. Sci.* **2020**, *10*, 5487. [CrossRef]
3. Nazeriye, M.; Haeri, A.; Martínez-Álvarez, F. Analysis of the Impact of Residential Property and Equipment on Building Energy Efficiency and Consumption—A Data Mining Approach. *Appl. Sci.* **2020**, *10*, 3589. [CrossRef]
4. Eroshenko, S.A.; Khalyasmaa, A.I.; Snegirev, D.A.; Dubailova, V.V.; Romanov, A.M.; Butusov, D.N. The Impact of Data Filtration on the Accuracy of Multiple Time-Domain Forecasting for Photovoltaic Power Plants Generation. *Appl. Sci.* **2020**, *10*, 8265. [CrossRef]
5. Trull, Ó.; García-Díaz, J.C.; Troncoso, A. Stability of Multiple Seasonal Holt-Winters Models Applied to Hourly Electricity Demand in Spain. *Appl. Sci.* **2020**, *10*, 2630. [CrossRef]

Article

The Impact of Data Filtration on the Accuracy of Multiple Time-Domain Forecasting for Photovoltaic Power Plants Generation

Stanislav A. Eroshenko [1,2], Alexandra I. Khalyasmaa [1,2], Denis A. Snegirev [1], Valeria V. Dubailova [1], Alexey M. Romanov [3] and Denis N. Butusov [4,*]

[1] Ural Power Engineering Institute, Ural Federal University named after the first President of Russia B.N. Yeltsin, 620002 Ekaterinburg, Russia; s.a.eroshenko@urfu.ru (S.A.E.); a.i.khaliasmaa@urfu.ru (A.I.K.); denis.snegirev@urfu.ru (D.A.S.); valeria.dubailova@urfu.ru (V.V.D.)

[2] Power Plants Department, Novosibirsk State Technical University, 630073 Novosibirsk, Russia

[3] Institute of Cybernetics, MIREA-Russian Technological University, 119454 Moscow, Russia; romanov@mirea.ru

[4] Youth Research Institute, Saint Petersburg Electrotechnical University "LETI", 197376 Saint Petersburg, Russia

* Correspondence: dnbutusov@etu.ru; Tel.: +7-950-008-7190

Received: 4 November 2020; Accepted: 20 November 2020; Published: 21 November 2020

Abstract: The paper reports the forecasting model for multiple time-domain photovoltaic power plants, developed in response to the necessity of bad weather days' accurate and robust power generation forecasting. We provide a brief description of the piloted short-term forecasting system and place under close scrutiny the main sources of photovoltaic power plants' generation forecasting errors. The effectiveness of the empirical approach versus unsupervised learning was investigated in application to source data filtration in order to improve the power generation forecasting accuracy for unstable weather conditions. The k-nearest neighbors' methodology was justified to be optimal for initial data filtration, based on the clusterization results, associated with peculiar weather and seasonal conditions. The photovoltaic power plants' forecasting accuracy improvement was further investigated for a one hour-ahead time-domain. It was proved that operational forecasting could be implemented based on the results of short-term day-ahead forecast mismatches predictions, which form the basis for multiple time-domain integrated forecasting tools. After a comparison of multiple time series forecasting approaches, operational forecasting was realized based on the second-order autoregression function and applied to short-term forecasting errors with the resulting accuracy of 87%. In the concluding part of the article the authors from the points of view of computational efficiency and scalability proposed the hardware system composition.

Keywords: photovoltaic power plant; short-term forecasting; data processing; data filtration; k-nearest neighbors; regression; autoregression

1. Introduction

Statistics show that photovoltaic power plants (PVPP) demonstrate the highest dynamics of installed capacity growth among renewable-based power plants worldwide [1]. However, given the climatic and geographical characteristics of Russian Federation territory, for the Unified Power System of Russia, as a whole, there is no considerable impact of stochastic renewable generation on the power system operation modes, but for the regional interconnected power systems (IPS) of the South and the Urals, a relatively high share of the installed capacity of PVPP is already observed. An increase of such power plants' share in the total power generation fleet leads to an even greater increase in their influence on the power balance [2], frequency [3,4], electric energy quality [5,6], static and dynamic stability [7–9], voltage levels [10,11] and electrical energy losses [12,13]. The first response to the PVPP installed capacity growth should be the development of forecasting systems, allowing reliable planning of power system operation modes.

In general terms power system planning is understood as an action schedule aimed at ensuring the balance of power consumption and generation, reliability and efficiency of the entire technological chain: power generation, transmission, distribution and consumption [14,15]. Short-term planning is typically carried out for the day-ahead perspective by the dispatch control centers. Operational planning is also carried out by dispatch centers, but within the operational day in order to sustain power balance and power supply reliability online.

When analyzing the renewable energy sources' (RES) influence on the power system operation mode, it is necessary to take into account the stochastic nature of weather conditions, especially for the PVPPs and wind power plants, power generation of which largely depends on meteorological factors [16,17]. This kind of uncertainty can be taken into account in their forecasting models both from the point of view of the probability theory [18] and the possibility theory [19]. When using the possibility theory, both qualitative and quantitative methods can be applied [20].

The PVPP generation forecasting is one of the most effective and least capital-intensive measures that allow the integration of stochastic generation sources into the power system and reduce the negative impact on the power system's operation mode. In this regard, the issue of day-ahead PVPP forecasting becomes a task with increasing priority in many countries [21].

Methods for PVPP generation forecasting can be divided into four main classes [22]: statistical, physical, intelligent and hybrid. Statistical models use statistical analysis to describe the relations between weather conditions and time series of solar irradiance or power generation of PVPPs, using retrospective data, as, for example, in [23–26]. Numerical weather forecast and satellite images, as suggested in the studies [27–30], form the basis for compiling PVPP generation forecasts for physical models. Intelligent models use artificial intelligence and machine learning methods to obtain forecasts of solar irradiance or PVPP power generation, mainly based on neural networks, as presented in [31–34]. Hybrid models for PVPP generation forecasting typically combine either physical or statistical or intelligent models. Examples of such models are presented in [35–38].

The authors of the article have previously developed their own step-by-step approach of PVPP generation short-term forecasting (STF), which is presented in detail in [39,40] and schematically given in Figure 1.

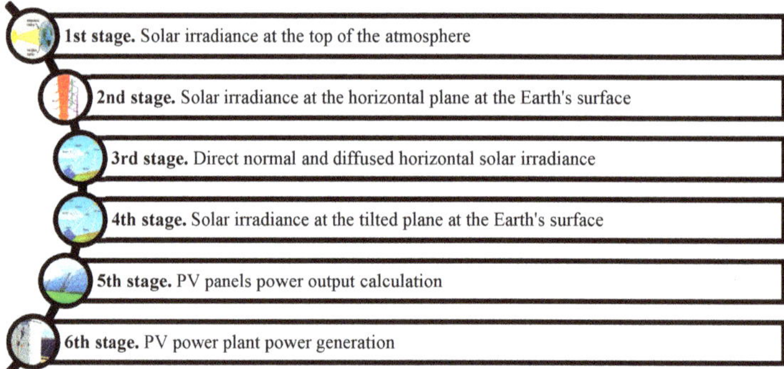

Figure 1. Flowchart of the photovoltaic power plants (PVPP) generation short-term forecasting algorithm.

The major advantage of the proposed approach is characterized by its flexibility in terms of accounting for the external factors, including the operational state of the power generation equipment, switchgear composition, operation modes of the adjacent power system, etc. due to the effective combination of solar irradiance forecasting algorithms with imitation model of the PVPP under consideration. Moreover, step-by-step calculation of PVPP generation provides extensive opportunities for the interpretation of forecasting results, unlike "black box" models, establishing correlation links between inputs (for instance, weather data) and output (PVPP generation). The PVPP short-term forecasting system was implemented in the industrial software package and piloted at the real PVPP allocated in one of the southern regions of the Russian Federation.

In general, in the course of piloting, the PVPP forecasting system demonstrated a satisfactory accuracy level for sunny and cloudy days [39,41]. However, the performance of the model significantly decreased during partly cloudy days and days with precipitation, which put the study on the model accuracy and robustness at the top of the priority list.

The present study focuses on the measures to be introduced to improve the PVPP forecasting system performance in terms of accuracy and robustness. The authors scrutinize the weather data analysis and highlight the actions to be taken in terms of source dataset processing to improve PVPP forecasting accuracy. It should be noted that since the source data processing is aimed at eliminating the noise (outliers, produced by extreme weather conditions), the proposed actions are applicable regardless of the initially applied forecasting methodology.

The second part of the study demonstrates the multiple time-domain hybrid approach, establishing the link between short-term PVPP generation forecast for the day-ahead perspective and operational forecast for intra-hour/intra-day planning of PVPP energy output. The latter one is implemented based on the supplementary function, characterizing the short-term forecast mismatch, giving the opportunity to evaluate the PVPP output for one hour-ahead time horizon, which resulted in integrated PVPP short-term and operational forecasting model.

The remainder of the article is organized as follows. In Section 2, the PVPP forecasting errors are analyzed and major error sources are highlighted. Section 3 presents the results of the studies on source data filtration to improve the forecasting accuracy and the model performance for unstable weather conditions. Section 4 introduces the concept of PVPP short-term forecast operational correction and implements PVPP operational forecast based on the short-term forecasting errors analysis. Section 6 describes the hardware required for the presented forecasting system implementation, focusing on computational performance and scalability. Finally, the Conclusion Section provides brief study outcomes.

2. The Main Sources of the PVPP Generation Forecasting Errors

Given the step-by-step procedure, introduced in the PVPP generation short-term forecasting algorithm, the total forecasting error of the PVPP generation is the sum of the errors of individual mathematical models, used at each step consequently. Moreover, there are special cases when the error of one of the models compensates for the error of the other one, but there are also scenarios when the errors of the models overlap each other, causing a significant increase in the total error.

As a rule, at the PVPP, there are measurements of the solar irradiance and the corresponding electrical energy generation at the alternating current side of the group of the inverters. Unfortunately, it is not possible to separately estimate the forecasting error that is introduced at each stage. However, these measurements allow for isolating the error components of the model, associated with tilted irradiance identification and PV panels and inverters outputs calculation (from 3 to 6 stages) and to estimate the total error of the forecasting methodology (from 1 to 6 stages). The calculation procedure uses data on the actual measurements of the solar irradiance acquired from the horizontally installed pyranometers. Figure 2 illustrates the forecasting error of the entire methodology on an hourly basis while using the actual metering data on solar irradiance. It is notable that the forecasted values of the PVPP generation are mainly conditioned by the forecasted values of the cloudiness, which were not observed at the given site. This also brings up an issue of the PVPP forecasting error estimation, including identification of the errors introduced by the data sources.

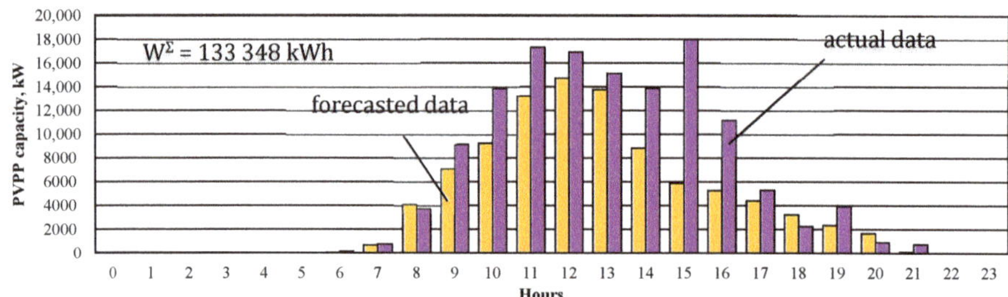

Figure 2. Actual versus Forecasted values of PVPP generation for day-ahead time horizon (stages 1–6).

Table 1 shows the calculation results for two PVPP forecasting scenarios and presents an estimate of the mean absolute percentage error (MAPE) without taking into account the error of stages 1–2 for the first scenario and taking into account the error of all stages for the second scenario.

Table 1. Errors for the daily forecast.

	Total Absolute Error, kW·h	Mean Absolute Percentage Error, %
Error of stages 3–6	11,590	8.7
Error of stages 1–6	43,112	32.3

The calculation results presented in Table 1 demonstrate that the main share of the PVPP forecasting error is introduced at the stage of calculating the global horizontal solar irradiance.

The value of the global horizontal solar irradiance, as well as the transparency index of the atmosphere, is largely determined by the cloudiness. The proportion of solar energy passing through the cloud layer is not constant. It depends on a several factors [40]:

- the numerical estimation of cloudiness (the proportion of the sky covered by clouds);
- type of clouds (cirrus, cumulus, stratus, etc.);
- cloud heights from the base to the top;
- the microstructure of clouds (e.g., water content per unit volume);
- distribution of clouds relative to the solar disk position, etc.;

In the majority of studies and for practical applications, given the limited amount of available meteorological data from local data providers, to assess the effect of cloudiness on the value of solar irradiance and the transparency index, as a rule, only one of the listed parameters is used—the cloudiness (in proportions or percentages)—since for this parameter it is the easiest to determine (to observe) and the easiest to forecast [41].

As a result, a situation arises when the parameter for which the forecast is made has an ambiguous effect on the forecasted value. Examples of such situations are described by the authors in [42,43]. As a result, at this stage of calculating the PVPP generation forecast, a significant forecasting error is introduced.

3. Data Filtering for Short-Term Forecasting of Photovoltaic Power Plants Generation

3.1. Case Study of the PVPP Generation Short-Term Forecast

The study object was a real PVPP located in the Astrakhan region of Russia, with an installed capacity of 15 MW. In total, 6076 observations are being examined for the period from 26 September 2017 to 05 February 2019. To verify the forecasting models, data for various characteristic periods were considered:

- spring weather period: 26 February 2018–11 March 2018, 14 days, 164 observations;
- summer weather period: 21 May 2018–1 June 2018, 12 days, 199 observations;
- autumn weather period: 11 September 2018–20 September 2018, 10 days, 130 observations;
- winter weather period: 28 January 2019–03 February 2019, 7 days, 81 observations.

The division into periods given in the upper list is intended to characterize the weather conditions corresponding to a particular season, not the calendar seasons. Therefore, those periods were chosen that corresponded to certain season from the point of view of weather conditions. In total, there are 574 observations in the studied data for a period of 43 days. Each observation includes the following information:

- *measured directly on site:*
    - actual PVPP generation, kWh;
    - actual global horizontal irradiance, W/m^2;
    - actual ambient temperature, °C;
    - actual wind speed, m/s;
- formed by the meteorological service:
    - actual and forecasted cloudiness, p.u.;
    - forecasted air temperature, °C;
    - forecasted wind speed, m/s;
    - actual and forecasted air humidity, p.u.

In addition, the calculations used passport and operational data of the PV panels and inverters.

3.2. Data Filtration Methods Application

One of the possible ways to improve the accuracy of calculating the transparency index can be the data sample filtering, which is used to calculate the coefficients of the regression model [40].

In order to study the possibility and evaluate the effectiveness of methods for improving the PVPP generation STF accuracy, a reference case study without data filtration is described in detail (Table 2, Figure 3).

Table 2 represents the numerical assessment of the forecasting quality without introducing data filtration approaches. The following denominations are used: W_Σ is the total PVPP energy production for the period under consideration, (kWh); E_Σ is the total absolute error for the period under consideration, (kWh); E_{avg} is the mean absolute error for the period under consideration, (kWh); σ_E is the absolute error standard deviation, (kWh); R^2 score is the determination coefficient, (p.u.); $SSEn$ is normalized sum of errors.

Table 2. Analysis of PVPP generation forecasting accuracy.

Parameter	26 February 2018–11 March 2018	21 May 2018–1 June 2018	11 September 2018–20 September 2018	28 January 2019–3 February 2019	For All Periods
W_Σ, kW·h	391,805.4	1,156,028.2	601,402.2	86,694.7	2,235,930.5
E_Σ, kW·h	185,497.2	253,367.0	258,279.6	64,829.7	761,973.5
E_{avg}, kW·h	1131.1	1306.0	1986.8	800.4	1339.1
σ_E, kW·h	1909.6	2201.9	2040.6	1303.4	2210.2
R^2					0.65
$SSEn$					14.77

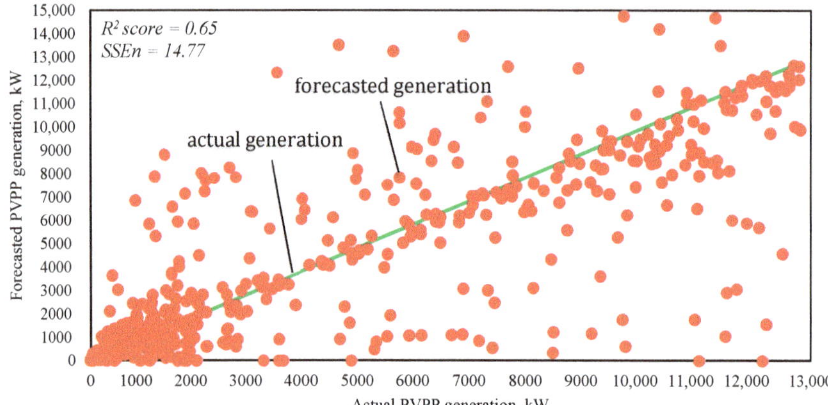

Figure 3. The scatter diagram of the PVPP generation forecasted values.

Due to the fact that when calculating SSE (sum of square error), very large values are obtained (for example, for the Figure 3 scenario the SSE would be equal to 3,322,661,719.61 kW^2), the $SSEn$ indicator is applied. It is calculated as the sum of squared errors normalized with respect to the square of the installed capacity of the photovoltaic power plant, $SSEn = SSE/(P_{inst})^2$, measured in p.u.

In Figure 3 there is a diagram showing the scatter of the forecasted values of PVPP generation relative to the actual values.

The determination coefficient characterizes a significant value of the PVPP generation forecasted values relative to the actual values. The reason for this is the cloudiness influence ambiguity on the value forecasted by the regression function—the transparency index.

Figure 4 shows an example of the transparency index dependence on cloudiness for the solar altitude angles, characterizing the morning and evening conditions ($\alpha < 15°$), as an example, and shows a significant uncertainty of the cloudiness influence cc on the transparency index k_T. For the same cloudiness value, the scatter in the transparency index values can reach 0.9 p.u.

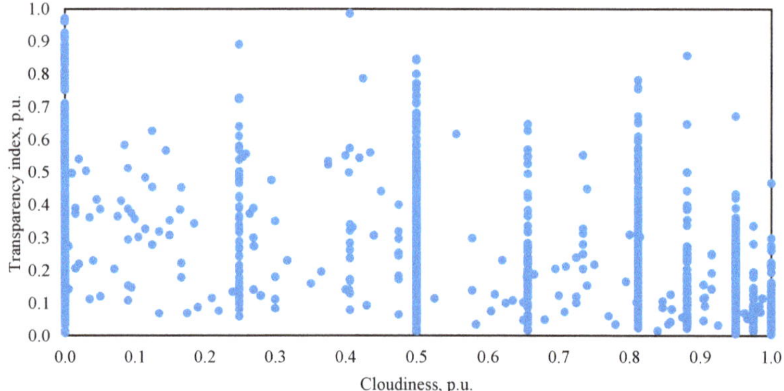

Figure 4. Transparency index versus cloudiness for morning/evening altitude angles.

3.3. Empirical Data Filtration

In determining outliers when analyzing the transparency index dependence on the cloudiness amount, an empirical formula can be used:

$$k_T \leq 0.9 - 0.5 \cdot cc, \tag{1}$$

where k_T—the transparency index, (p.u.); cc—the cloudiness, (p.u.).

This expression is determined on the assumptions that come from the experience of PVPP solar forecasting system application at the real power generation facility: under absolutely cloudless weather conditions the transparency index value k_T cannot exceed 0.9 p.u., and under the most cloudy weather the transparency index value k_T cannot exceed 0.5 p.u.

All observations above the straight line $k_T = 0.9 - 0.5 \cdot cc$, are treated as outliers and are considered unreliable. Figure 5 shows the transparency index dependencies on cloudiness, filtered in accordance with the expression (1), for the above-described ranges of the solar altitude angle, characterizing the morning and evening conditions ($\alpha < 15°$).

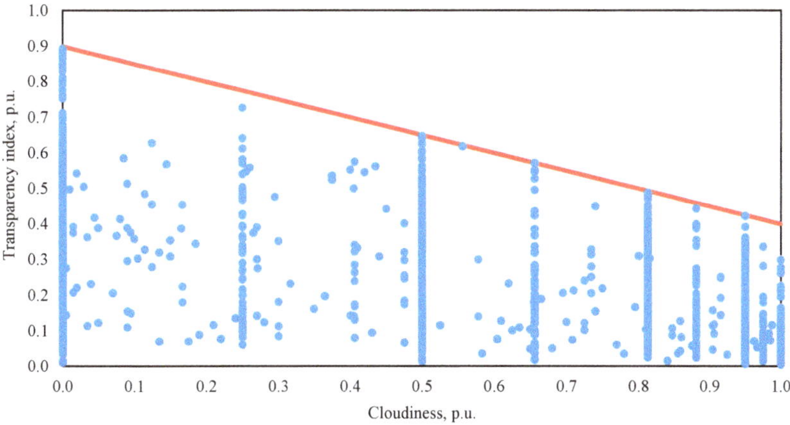

Figure 5. Transparency index versus cloudiness for morning/evening altitude angles with empirical filtration.

From Figure 5 it appears an uncertainty decrease in the cloudiness influence cc on the transparency index k_T due to the introduced filtering of the observations. The results of PVPP generation forecasting accuracy assessment using empirical filtering are shown in Figure 6.

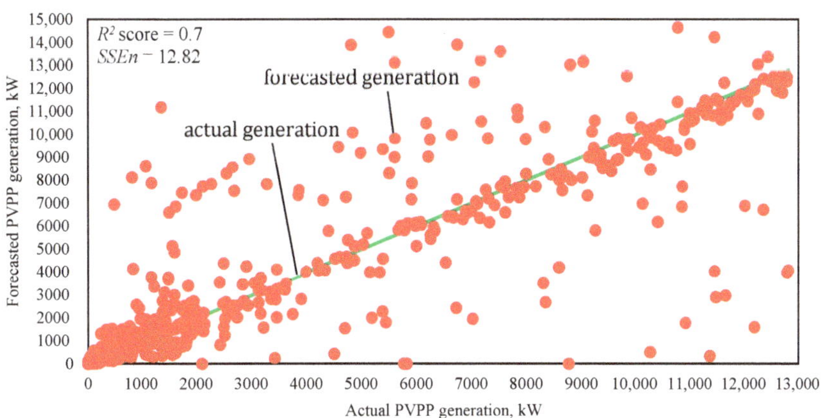

Figure 6. The scatter diagram of the PVPP generation forecasted values with empirical filtering.

The comparison of Figures 3 and 6 allows one to evaluate the efficiency of using simple filtration. The total error value for all the specific periods under consideration using an empirical filter is lower than for the calculations without using the filter, which is confirmed by the scatter on the diagrams. The determination coefficient R^2 has also increased.

The advantages of the empirical filtering method are the usability and the fewer required computing resources. Disadvantages of the filtering method are as follows: the difficulty to accurately identify outliers of the transparency index k_T, the excessive observations filtering (in addition to outliers, reliable observations can be discarded) and the lack of empirical expressions versatility.

3.4. The K-Means Filtration Method

In order to more accurately identify outliers, the authors of the study used the k-means method. The k-means method is implemented in accordance with the following algorithm [42], as shown in Figure 7 for 200 randomly generated observations and 6 cluster centers for two-dimensional space (the number of features describing each observation is 2). The step-by-step procedure of the methodology is given in Figure 8. A silhouette measure is used to assess the quality of data clustering.

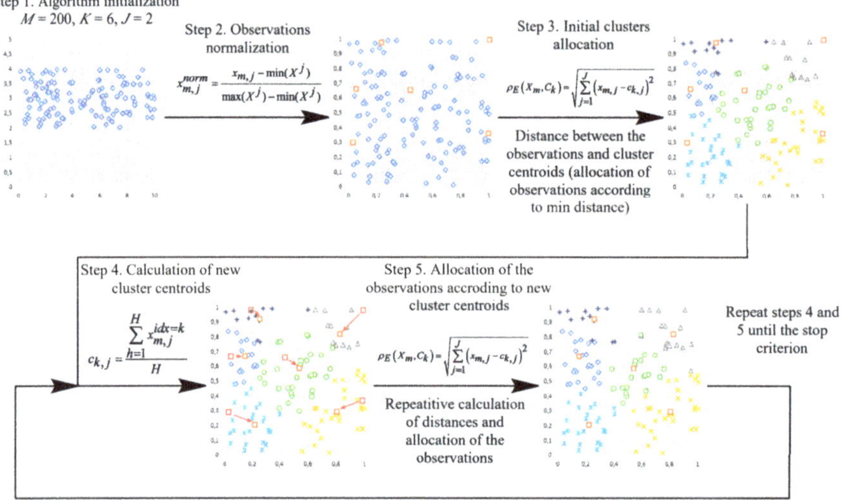

Figure 7. K-means based data clustering.

The silhouette measure characterizes the distance of the observation from the nearest cluster to which it does not belong. The silhouette measure is determined in accordance with the expression:

$$Sil_m = \frac{B - A}{\max(A, B)} \quad (2)$$

where Sil_m—the silhouette measure for the observation; A—the distance from the observation to the nearest cluster center, to which this observation belongs; B—the distance from the observation to the nearest cluster center to which this observation does not belong.

By the silhouette measure value, the division quality is determined: poor division quality is characterized by the measure values from −1 to 0.2; middle division quality—from 0.2 to 0.5; good division quality—from 0.5 to 1.

To identify the transparency index k_T anomalous behavior and to solve the problem of more accurate outliers identification, a three-dimensional feature space was used in the study, based on the k-means clustering method. 6076 observations were analyzed for the period from 26/09/17 to 05/02/19. Cloudiness cc, solar altitude angle sine $\sin\alpha$ and solar inclination angle δ were used as features describing the observations.

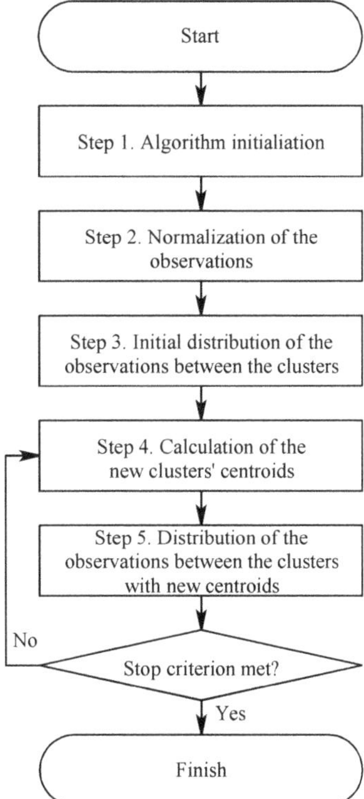

Figure 8. Flow-chart of the algorithm.

Solar altitude angle α is determined according to the expression below:

$$\alpha = 90 - \arccos(\cos\varphi\cos\delta\cos\omega + \sin\varphi\sin\delta), \quad (3)$$

where φ—the latitude, (deg); δ—the solar inclination angle, (deg); ω—the solar hourly angle, (deg).

To calculate the solar inclination angle, the following expression is used:

$$\delta = 23.45^\circ \sin\left(360^\circ \frac{284+n}{365}\right), \quad (4)$$

where n is the day number.

To select the number of clusters and determine the initial approximations of the cluster centers, a combination of various characteristic values of each feature was used. The list of characteristic values and the corresponding descriptions are presented in Table 3.

Table 3. Description and features values for clusters initial approximations.

Solar Altitude Angle Sine		**Cloudiness**		**Solar Inclination Angle**	
Description	*Value*	*Description*	*Value*	*Description*	*Value*
morning/evening	0.1	almost no clouds	0.1	winter	0.1
late night/day	0.3	low clouds	0.5	off-season closer to winter	0.33
late morning	0.5	medium clouds	0.8	off-season closer to summer	0.66
midday	0.7	heavy clouds	1	summer	1

According to Table 3, various combinations of cloudiness values, solar altitude angle sine and solar inclination angle are formed to determine the initial approximations of the cluster centers.

The number of possible combinations is 43. Since in the range of the solar altitude angle sine close to 0.7 (midday), there are no points with coordinates along the axis δ in the range of 0.1–0.33 (winter-off-season closer to winter), combinations with such values are not included in the final set of cluster centers' initial approximations.

A total set of 56 different combinations is formed. Figure 9a,b illustrate the location of 6076 observations (blue markers) and 56 initial approximations of cluster centers (black markers) in a three-dimensional feature space $\sin\alpha - cc - \delta$. The clustering observations results using the k-means method are presented in Figure 9c,d, as well as in Table 4. The value of the final silhouette measure obtained from the clustering results by the k-means method, averaged between all observations, is 0.52, which confirms the good quality of the observations division into clusters. The calculation time was 1 h 56 min.

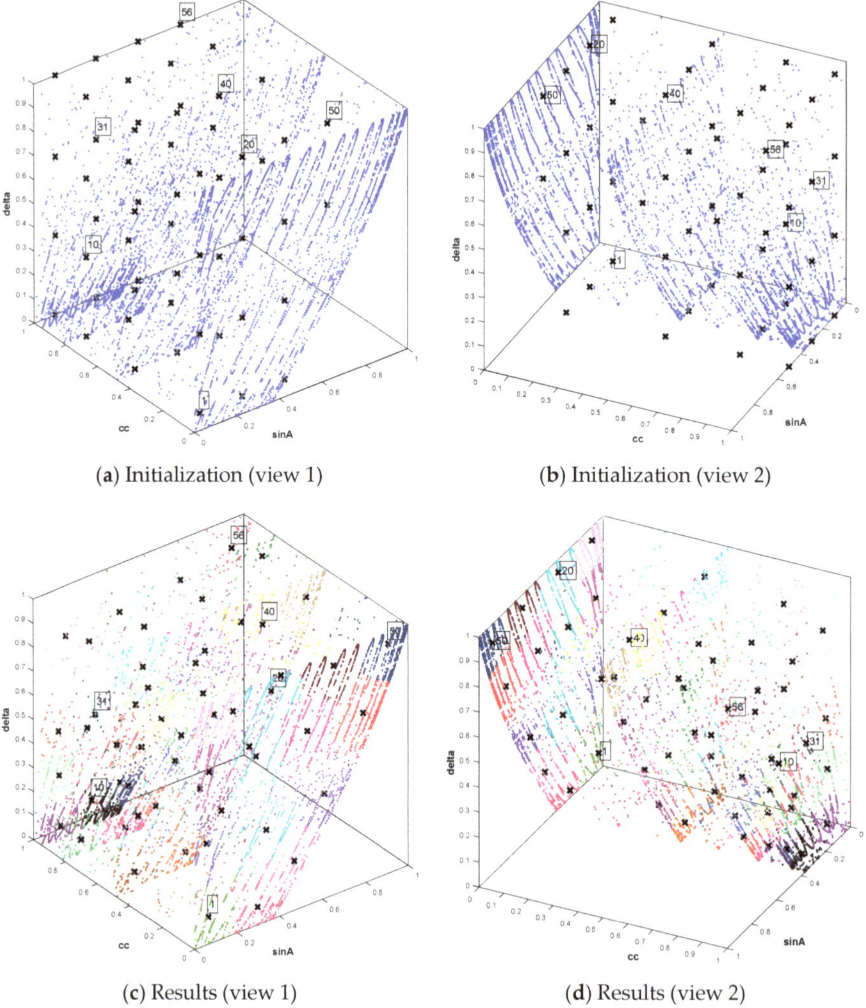

(**a**) Initialization (view 1) (**b**) Initialization (view 2)

(**c**) Results (view 1) (**d**) Results (view 2)

Figure 9. Geometric interpretation of data initialization and clustering results using the k-means method.

Table 4. Results of calculating the parameters for forecast accuracy estimating for various combinations.

$X - n \cdot \sigma_{k_T} \leq k_T \leq X + n \cdot \sigma_{k_T}$	E_Σ, kW·h	σ_E, kW·h	E_{avg}, kW·h	$E_\Sigma^{\%}$, %
X—arithmetic mean, $n = 0.5$	464,104.24	1560.80	815.65	20.76
X—arithmetic mean, $n = 1.0$	442,004.04	1486.48	776.81	19.77
X—arithmetic mean, $n = 1.5$	437,334.47	1274.05	768.60	19.56
X—arithmetic mean, $n = 2.0$	496,591.54	1470.06	872.74	22.21
X—median, $n = 0.5$	449,945.86	1361.23	790.77	20.12
X—median, $n = 1.0$	424,117.13	1271.61	745.37	18.97
X—median, $n = 1.5$	417,287.68	1214.58	733.37	18.66
X—median, $n = 2.0$	438,152.06	1289.76	770.04	19.60
X—mode, $n = 0.5$	495,570.51	1466.62	870.95	22.16
X—mode, $n = 1.0$	466,985.74	1373.99	820.71	20.89
X—mode, $n = 1.5$	471,971.92	1387.26	829.48	21.11
X—mode, $n = 2.0$	530,260.45	1583.29	931.92	23.72

The k-means clustering makes it possible to divide observations into clusters that are similar in terms of cloudiness, time of day and season. Observations combined in this way should have similar transparency index values k_T.

A number of parameters are used for the numerical evaluation of transparency index observations obtained within the clusters: arithmetic mean value, median, mode, mean-square deviation. Observations that differ significantly from the rest in the cluster are recognized as outliers and are considered unreliable. The following expression is used for determining the cluster outliers:

$$X - n \cdot \sigma_{k_T} \leq k_T \leq X + n \cdot \sigma_{k_T}, \tag{5}$$

where k_T—the transparency index, (p.u.); σ_{k_T}—the transparency index standard deviation, (p.u.); n—the number of transparency index standard deviations; X—arithmetic mean, median or mode of the transparency index for a given cluster (p.u.).

Table 4 shows the results of calculating the parameters for forecasting accuracy estimation for various combinations of X and n.

The analysis of the Table 4 shows that the observations filtering using the transparency index median of each cluster has the greatest efficiency in determining outliers within the clusters; the number of the transparency index mean-square deviations for the confidence range is 1.5. The results of calculating the parameters for estimating the PVPP generation forecasting accuracy using the k-means method for filtering data are presented in Table 5.

Table 5. Results of calculating the parameters for estimating the PVPP generation forecasting accuracy using the k-means method for data filtering.

Parameter	26 February 2018–11 March 2018	21 May 2018–1 June 2018	11 September 2018–20 September 2018	28 January 2019–3 February 2019	For All Periods
W_Σ, kW·h	391,805.4	1,156,028.2	601,402.2	86,694.7	2,235,930.5
E_Σ, kW·h	113,424.0	75,093.5	184,327.4	45,661.1	417,287.7
E_{avg}, kW·h	691.6	377.4	1417.9	563.7	733.4
σ_E, kW·h	997.7	893.1	1724.2	733.3	1 214.6
$E_\Sigma^{\%}$, %	29.0	6.50	30.7	52.7	18.7
R^2					0.88

Advantages of filtering the initial data using the k-means method: more accurate transparency index outliers identification compared to an empirical model, since the analysis takes into account the season and day time; the versatility of the proposed method for power plants various geographic locations; less chance of over-filtering observations compared to the empiric approach.

Disadvantages of filtering the initial data using the k-means method: high costs of computing resources, the calculation time for the considered example reached almost two h; more complex algorithm compared to the empirical model.

3.5. Filtration Models Comparative Analysis

Table 6 shows the comparison of the parameters for assessing the PVPP generation forecast accuracy for three scenarios: without filtering the source data, using an empirical filter, introducing the k-means approach.

Table 6. Comparison of the short-term forecasting (STF) results for various filtration models.

Parameter	Without Filtration	Simple Filter	K-Means Method
W_{Σ}, kW·h	2,235,930.48	2,235,930.48	2,235,930.48
E_{Σ}, kW·h	761,973.46	640,411.96	417,287.68
E_{avg}, kW·h	1339.14	1151.29	733.37
σ_E, kW·h	1909.59	2242.59	1214.58
$E_{\Sigma}^{\%}$, %	34.08	28.64	18.66
R^2	0.65	0.70	0.88

The error assessment criteria analysis shows that the observations filtering using the k-means method has the best performance. The total error value is reduced by almost 2 times compared with the calculation without filtering and more than 1.5 times compared with the calculation using an empirical filter.

4. Photovoltaic Power Plants Generation Short-Term Forecast Operational Correction

4.1. General Approach to the PVPP Generation Operational Forecasting Models

When addressing the problem of PVPP short-term forecasting accuracy improvement, it should be outlined that there are several fundamentally different degrees of latitude. Typically, at first, the investigators justify the particular types and parameters of the forecasting approach. The next step is often addressing the data analytics issues, including Feature Engineering, applying practical knowledge to the dataset processing, data gaps elimination, outlier filtration, etc. The last, but not the least, direction to minimize the PVPP generation forecasting error is an adjustment of time resolution of the model. Indeed, it is naturally evident, that very-short term forecasts of PVPP generation demonstrate more accurate results for intra-hour or intra-day periods than multiple day-ahead forecasting models, based on numerical weather predictions. However, typically, the investigators address the behavior of their approaches for static time-domain models, which is justified in the majority of cases by the necessity to introduce another mathematical basis for different time-domains or other structures of the mathematical core of the proposed models, other features with different time resolutions, etc. The static time-domain operational forecasts of PVPP generation for an hour ahead perspective often turns out to be problematic, since the hourly interval is founded to be too large for the models with smaller time resolution (1-min, 5-min, 15-min, etc.) as far as one hour-ahead calculations in such circumstances are to be treated as 60, 12, 4 periods ahead forecasts, respectively. For this reason, the correlation between the PVPP generation at two adjacent hourly intervals is often poorly traced. Furthermore, vice versa, using single hour resolution models for hour-ahead perspective does meet the requirements of the power system operational control since the intra-hour deviations of PVPP generation are not taken into account.

In the present study, the authors have attempted to establish the bridge between the STF day-ahead forecasting system, implemented on multiple regression with k-means initial dataset filtration, with the operational hour-ahead PVPP generation forecast. The latter one is implemented on the basis of the supplementary STF error forecasting function, giving the opportunity to evaluate the STF error for the hour-ahead time horizon. The knowledge on what would be the mismatch of the STF for the hour-ahead perspective gives the opportunity to implement an operational (very-short term) forecast on the basis of the initially developed and optimized STF approach by providing STF forecasting error correction. The proposed approach is justified by another fact that PVPP generation STF error, if it occurs, exists for several time intervals straight, that is, several hours. This circumstance initiates using retrospective STF error data to make operational forecasts, since the STF errors time series turns out to be more predictable than of the PVPP operational forecast one because of the data noise, appearing in smaller time-domain models.

The object of study for operational forecasting is the same PVPP, as was investigated for the STF. Within this study, different methodologies of calculating the forecast for 1 h ahead are considered. In order to implement the operational forecast, based on the retrospective data of the STF errors, an STF error forecast is calculated for an hour ahead horizon. Based on the calculation results, the STF is corrected for the STF forecasted error, which will be essentially the operational forecast.

Thus, when compiling the operational forecast in all the models, the STF error appears as the forecasted value when determining the one-hour-ahead PVPP generation:

$$E_{\text{stf}}^{\text{act}} = W_{\text{act}} - W_{\text{stf}}, \tag{6}$$

where $E_{\text{stf}}^{\text{act}}$—PVPP generation STF error, (kW·h); W_{act}—PVPP generation actual value, (kW·h); W_{stf}—PVPP generation STF, (kW·h).

The calculation algorithm used for operational PVPP generation forecasting consists of the following items, presented in Figure 10, where W_{of}—PVPP generation operational forecast, (kW·h); W_{stf}—PVPP generation STF, (kW·h); $E_{\text{stf}}^{\text{f}}$—operational forecast of PVPP generation STF error, (kW·h).

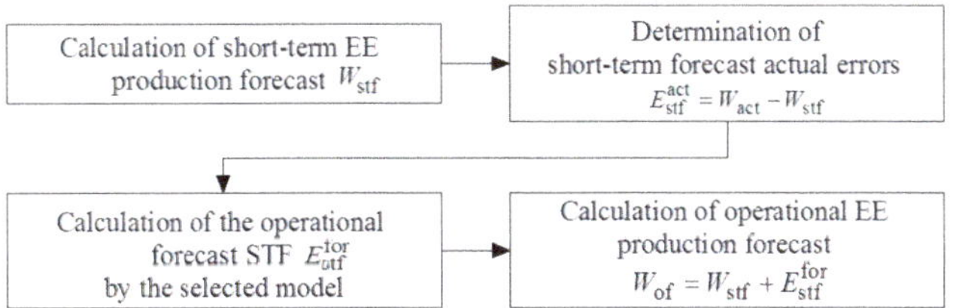

Figure 10. Block diagram of operational forecast algorithm.

The proposed algorithm makes it possible to promptly (1 h ahead) correct STF stationary errors (that is, those that occur for several hours straight). In this case, the cumulative error of the methodology for calculating the STF will be corrected: the error associated with the assessment of the cloudiness influence on the share of solar energy losses when passing through the cloud layer; errors in cloudiness forecasts, as well as errors of other mathematical models.

4.2. PVPP Generation Operational Forecasting Models Description

To implement the operational forecasting on the basis of retrospective data on STF errors, the possibility of using a number of statistical mathematical models is considered [43].

4.2.1. Persistence Model (Represents the So-Called "Naive" Approach)

According to this approach, it is assumed that the forecasted value at the next time step is equal to the actual value at the current step. Thus:

$$E^{\mathrm{pr}}_{\mathrm{stf}}(t+1) = E^{\mathrm{act}}_{\mathrm{stf}}(t), \tag{7}$$

where t—time interval, (hour); $E^{\mathrm{pr}}_{\mathrm{stf}}(t+1)$—STF error operational forecast for 1 h ahead, (kW·h); $E^{\mathrm{act}}_{\mathrm{stf}}(t)$—actual value of the STF error, (kW·h).

This model makes it possible to obtain fairly accurate transparency index operational forecasts in those cases when a stationary STF error occurs, which persists for several time intervals straight, constant in magnitude and sign. At the same time, in the case of a rapid change in the STF error value or sign, the operational forecast will be less accurate than the STF.

4.2.2. Moving Average Model (Is an Advanced Inertial Model)

This is a well-known time series smoothing technique that eliminates random fluctuations in the time series. The moving average model (MA (M)) can be represented in accordance with the following expression:

$$E^{\mathrm{pr}}_{\mathrm{stf}}(t+1) = \frac{1}{T}\sum_{i=0}^{T-1} E^{\mathrm{act}}_{\mathrm{stf}}(t-i), \tag{8}$$

where T—the number of observations in the period used to calculate the mean, (dimensionless value); i—the offset relative to the current time interval, (hour); $E^{\mathrm{act}}_{\mathrm{stf}}(t-i)$—STF error actual value for the time interval $t-i$, (p.u.), the number of observations in the period—T, used to calculate the average value, denotes the model order.

When using the MA (M) model, the initial values of the time series are replaced by the arithmetic mean within the selected time period. When forecasting for the next interval, the period is shifted by one observation, and the calculation of the mean is repeated. The periods use the same time frames for determining the average. The wider the frame used for smoothing, the smoother the trend is.

4.2.3. Autoregressive Model

Time series model (AR (p)), where the time series values are linearly dependent on the previous values of the same series. It is assumed that STF errors time series can be represented as an autoregressive function, since this random process proceeds approximately uniformly in time, while random fluctuations occur around some mean value close to zero. Moreover, neither the average amplitude nor the nature of these fluctuations show significant changes over time. The autoregressive process is defined as follows:

$$E^{\mathrm{pr}}_{\mathrm{stf}}(t+1) = c + \sum_{i=1}^{p} a_i \cdot E^{\mathrm{act}}_{\mathrm{stf}}(t-i), \tag{9}$$

where c—the constant (free) term, (dimensionless quantity); p—the model order, that is the number of previous time intervals used for the calculation, (dimensionless value); a_i—the autoregressive coefficient for $t-i$ time interval, (dimensionless quantity).

To estimate the autoregression coefficients, as well as for other regression models, the least squares method can be used. The autoregression coefficients are calculated using the Gauss transformation:

$$A = \left(X^T X\right)^{-1} X^T Y, \tag{10}$$

where A—the autoregression coefficients vector; X—the independent variables matrix, composed of actual values $E^{\mathrm{act}}_{\mathrm{stf}}(t-i)$; Y—the dependent variables vector, composed of actual values $E^{\mathrm{act}}_{\mathrm{stf}}(t)$.

4.2.4. Autoregressive Moving Average Model

The ARMA(p,T) model is a generalization of MA иAR processes. The STF errors time series treatment, which was smoothed using the moving average model, shows that it is stationary, as the original time series. Thus, it is also possible to use an autoregressive model for the moving averages time series. The autoregressive process for a moving average is defined as follows:

$$E^{pr}_{stf}(t+1) = c + \sum_{i=1}^{p} a_i \cdot \left[\frac{1}{T} \sum_{j=0}^{T-1} E^{act}_{stf}(t-j) \right], \tag{11}$$

where p—the model order, that is, the number of the moving average previous values used for the calculation, (dimensionless value); a_i—the autoregressive coefficient for $t-i$ the moving average value, (p.u.); i—the shift relative to the time interval corresponding to the moving average current value, (hour); j—the shift relative to the current time interval, (hour); T—the number of observations in the period used to calculate the average, (dimensionless value); $E^{act}_{stf}(t-j)$—the STF error actual value for the time interval $t-j$ within the period T, (kW·h), where the number of observations in the period T, used to calculate the average denotes the order of the MA model, and the number of moving averages p, used for the calculation denotes the order of the AR model.

4.2.5. Autoregressive Model with Exogenous Inputs

ARX(p,q) model represents an autoregressive process that also takes into account the values that do not belong to the considered time series. As previously noted, a simple autoregressive AR model uses the previous time series values as features, without using any other features. At the same time, when making the forecast, one may take into account other features that affect the transparency index, that is, the cloudiness cc. The autoregressive process using cloudiness cc as an exogenous feature is defined as follows:

$$E^{pr}_{stf}(t+1) = c + \sum_{i=1}^{p} a_i \cdot E^{act}_{stf}(t-i) + \sum_{i=0}^{q} b_i \cdot cc(t-i), \tag{12}$$

where p—the order of the autoregressive inputs, that is the number of STF error previous values used for the calculation, (dimensionless value); a_i—the autoregression coefficient for $t-i$ of STF error value, (dimensionless value); q—the exogenous inputs order, that is, the number of cloudiness values used for the calculation, (dimensionless value); b_i—the autoregression coefficient for $t-i$ of cloudiness, (dimensionless value); $cc(t-i)$ the actual cloudiness value for the time interval $t-i$, (p.u.); the number of the STF error previous values p, used for the calculation denotes the AR model order, and the number of cloud values q, used for the calculation denotes the X model order.

5. Comparison of Operational and Short-Term Forecasting Models

Table 7 shows the parameters comparison for assessing the PVPP generation forecast accuracy for various operational forecast models. As far as the models under consideration have different number of predictors, the adjusted R^2 metrics is introduced as a model quality criterion. As in the previous case (short-term forecast), adjusted R^2 also demonstrates how well the model fits the data, but the total score is adjusted according to the number of terms in the model:

$$R^2_{adjusted} = 1 - \left(1 - R^2\right) \times (k-1)/(k-n-1), \tag{13}$$

where R^2—R-square measure; n—number of samples in the set; k—number of variables in the model.

The error assessment analysis shows that the second-order autoregressive model AR(2) has the best indicators among the operational forecast models. The parameters characterizing the results of PVPP generation operational forecast accuracy assessment using the second-order autoregressive model AR(2) are presented in Table 8.

Table 7. Parameters comparison for PVPP operational forecast accuracy assessment.

Model	Parameters				
	E_{Σ}, kW·h	E_{avg}, kW·h	σ_E, kW·h	$E_{\Sigma}^{\%}$, %	$R^2_{adjusted}$
Model P	346,892.60	1063.18	609.65	15.51	0.874
Model $MA(2)$	337,843.22	1035.45	593.75	15.11	0.887
Model $MA(3)$	347,495.89	1065.03	610.71	15.54	0.874
Model $MA(4)$	355,941.97	1090.92	625.56	15.92	0.862
Model $MA(5)$	364,991.34	1118.65	641.46	16.32	0.851
Model $AR(1)$	325,777.39	998.47	572.54	14.57	0.904
Model $AR(2)$	301,645.74	924.51	530.13	13.49	0.940
Model $AR(3)$	331,810.31	1016.96	583.15	14.84	0.895
Model $ARMA(1,2)$	328,793.85	1007.71	577.85	14.71	0.900
Model $ARMA(2,2)$	337,843.22	1035.45	593.75	15.11	0.887
Model $ARMA(3,2)$	358,958.43	1100.16	630.86	16.05	0.858
Model $ARMA(2,3)$	352,925.51	1081.67	620.26	15.78	0.866
Model $ARX(2,1)$	327,180.05	1002.77	575.01	14.63	0.902
Model $ARX(2,2)$	330,196.50	1012.01	580.31	14.77	0.898
Model $ARX(2,3)$	333,212.96	1021.26	585.61	14.90	0.893

Table 8. The parameters calculating results for assessing the PVPP generation operational forecast accuracy.

Parameter	26 February 2018–11 March 2018	21 May 2018–1 June 2018	11 September 2018–20 September 2018	28 January 2019–3 February 2019	For All Periods
W_{Σ}, kW·h	391,805.4	1,156,028.2	601,402.2	86,694.7	2,235,930.5
E_{Σ}, kW·h	82,951.0	76,503.8	119,241.6	23,406.3	301,645.8
E_{avg}, kW·h	505.8	384.4	917.2	289.0	530.1
σ_E, kW·h	722.5	867.6	1011.8	440.0	924.5
$E_{\Sigma}^{\%}$, %	21.17	6.62	19.83	27.00	13.5
R^2					0.94

Figure 11 demonstrates the comparison of the PVPP generation actual values, the PVPP generation STF using the initial data filtration by the k-means method and the operational forecast using the second-order autoregressive model AR(2).

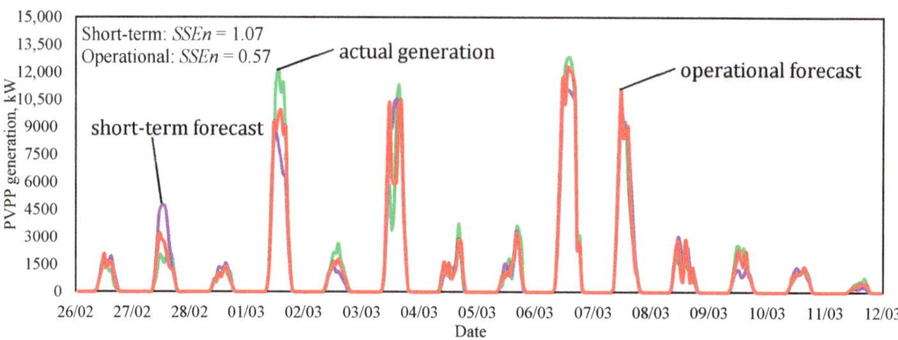

(**a**) Calculation results for 26 February 2018–12 March 2018.

Figure 11. *Cont.*

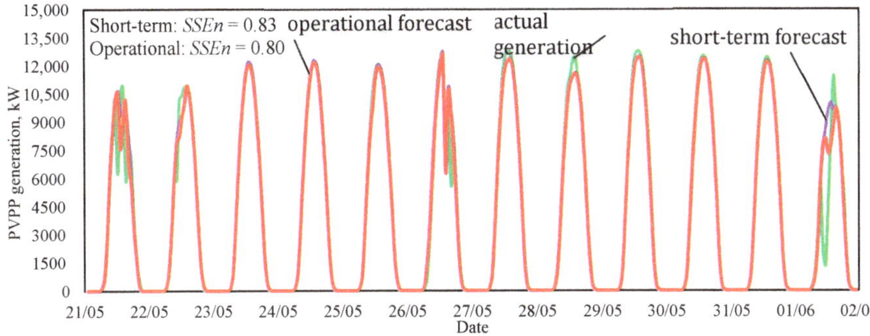

(b) Calculation results for 21 May 2018–2 June 2018.

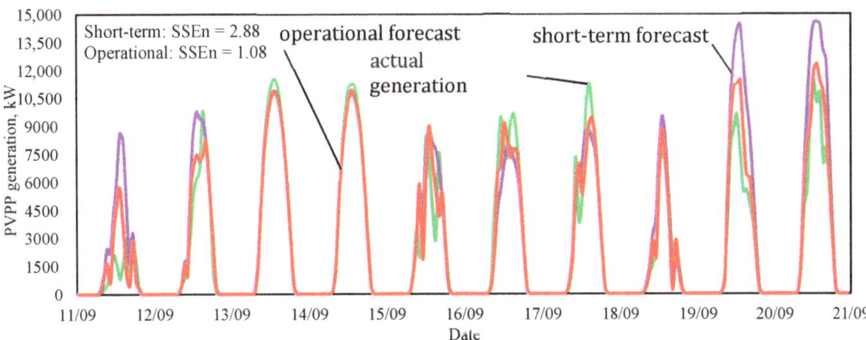

(c) Calculation results for 11 September 2018–21 September 2018.

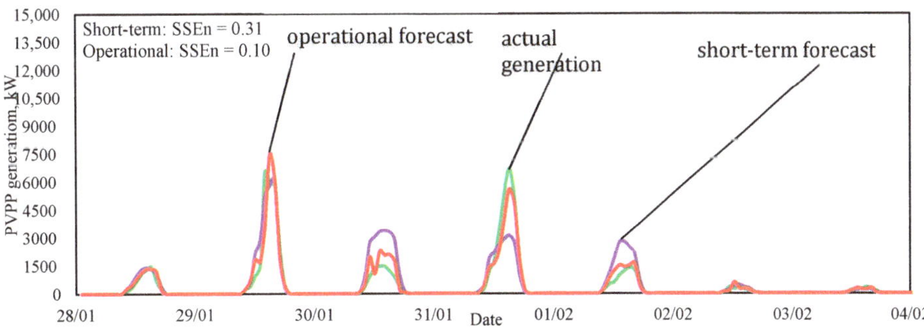

(d) Calculation results for 28 January 2019–5 February 2019.

Figure 11. The comparison of short-term and operational PVPP forecasts with actual values.

As one can see from Figure 11a–d, the operational forecast makes it possible to refine the short-term forecast, the curve of operational forecast absolute errors has a flat character compared to the short-term forecast, the absolute forecasting error values are closer to zero. Table 9 demonstrates the comparison of the short-term and operational PVPP generation forecasting accuracy.

When dealing with stochastic phenomena, providing a 100% accurate forecast is problematic. In most cases, we can only talk about the probability of the forecasted to meet the confidence range. To assess the reliability of the proposed short-term and operational forecasting models, the confidence probabilities corresponding to the intervals of ±1 MW and ±2 MW were also calculated. The values of these probabilities are presented in Table 10.

Table 9. The parameters comparison for assessing the short-term and operational PVPP generation forecast accuracy.

Parameter	STF, *K*-Means Methodology	Operational Forecast, AR(2)
W_Σ, kW·h	2,235,930.48	2,235,930.48
E_Σ, kW·h	417,287.68	301,645.74
E_{avg}, kW·h	733.37	530.13
σ_E, kW·h	1214.58	924.51
$E_\Sigma^{\%}$, %	18.66	13.49
R^2	0.88	0.94

Table 10. The reliability parameters comparison of PVPP generation operational and short-term forecast.

Confidence Interval	The Share of the Forecasts within the Interval	
	STF, *K*-Means Method	Operational Forecast, AR(2)
±1 MW (6.7% from PVPP P_{inst})	78.9%	88.5%
±2 MW (13.3% from PVPP P_{inst})	92.2%	97.5%

As can be seen from Table 10, for STF 78.9% of forecasts are characterized by an error not exceeding 1 MW (6.7% of the plant's installed capacity), 92.2% of forecasts are characterized by an error not exceeding 2 MW (13.3% of the plant's installed capacity). At the same time, for operational forecast, 88.5% of the forecasts are characterized by an error not exceeding 1 MW, and 97.5% of forecasts are characterized by an error not exceeding 2 MW.

The results of the parameter comparison for assessing the PVPP generation short-term and operational forecasts show that the error characteristics are finally improved by almost 1.5 times. At the same time, the operational forecast makes it possible to obtain more reliable estimates of the PVPP generation forecasted values: the number of operational forecasts belonging to the confidence interval of ±1 MW is almost 10% more, and the number of forecasts belonging to the confidence interval of ±2 MW is observed 5% more frequently. Thus, we can conclude about the effectiveness of the proposed methodology for operational forecasting based on retrospective data on short-term forecasting errors.

6. Forecasting System Hard-Ware Implementation for Autonomous Photovoltaic Power Plants

The discussed above forecasting algorithms were implemented using a distributed computing system that included a low-cost x86 embedded computer and an FPGA (Figure 12).

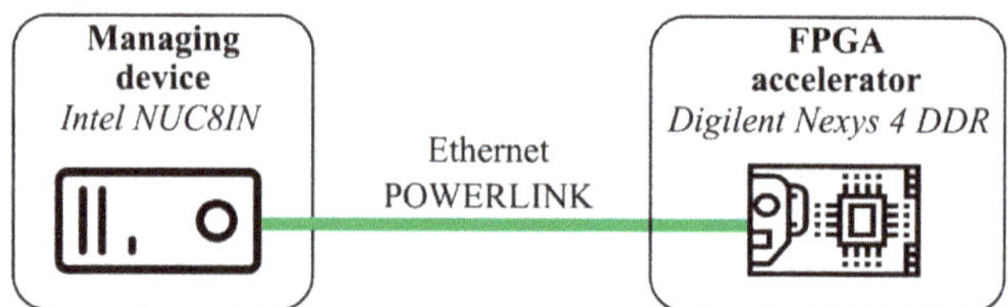

Figure 12. Hardware architecture used for forecasting algorithms implementation.

In general, the architecture of the implemented computing system repeats the one proposed in [44]. FPGA in this architecture is used as a custom accelerator, which is built on a basis of one of more specialized computational cores (SCCs) and Ethernet POWERLINK communication core [45]. Each SCC, which includes a combination of a matrix coprocessor [46] and a general-purpose processor and performs computations requested by managing device, built on the basis of an x86 embedded computer running Linux operating system. This architecture is inspired by the ones used in space robotics [47], which makes it suitable and attractive for adaptation to other application areas where the robust and maintenance-free operation of equipment is required in long terms; for example, for autonomous PVPPs.

The developed x86 software algorithms and FPGA firmware were first verified using the Vmodel toolbox [48] simulation tool, and then uploaded into Intel NUC8IN computer and a Digillent Nexys 4 DDR kit FPGA (Figure 12). The estimation evaluated during model verification was the same as the ones provided by the real hardware. The worst-case difference between evaluated forecasts and reference ones presented in Table 7 was 0.04883%. This error is caused by the use of fixed-point calculations on FPGA accelerators. Meanwhile, as it can be seen from Table 7, it is significantly less than the error of the forecasting methods themselves.

This experiment demonstrates that the discussed forecasting methods can be implemented using embedded equipment and integrated into autonomous photovoltaic power plants. Moreover, the proposed implementation approach is scalable and, as it was shown in [44], even a significantly larger amount of data can be processed without the use of high-performance servers just by increasing the number of distributed FPGA-based accelerators.

7. Conclusions

The present study addresses the development of short-term and operational forecasting methods of photovoltaic power plants, which arose in response to the energy sector transition, characterized by the growing share of stochastic power generation. As it was stated in the introduction, reliable forecasting systems are the most effective and least capital-intensive measures that allow for the integration of stochastic generation sources into the power systems.

Based on the results of the previously carried out investigations, a short-term day-ahead PVPP generation forecasting system was developed by an effective combination of astronomical and statistical approaches. The step-by-step forecasting procedure, comprising the global horizontal irradiance identification, tilted surface irradiance assessment and, finally, calculation of the PV power plant output was implemented, providing rich opportunities for forecasting results' interpretation and overall model flexibility, allowing us to take into account the operational state of the power generation equipment, switchgear composition, operation modes of the adjacent power system and other external factors.

In the course of the short-term forecasting system industrial piloting, it was determined that the greatest error (more than 70% of the total value of mean absolute percentage error) in the PVPP generation forecast calculation is introduced at the stage of determining the transparency index according to the regression model due to the ambiguity of cloudiness impact on the forecasted value of the transparency index. Given that from the point of view of PVPP power output, there is a fundamental difference of various meteorological conditions and events, which may be characterized by similar weather forecasting parameters, acquired from the open-source weather data provider, it was decided to focus the attention on the dataset filtration. The idea was to eliminate the dataset outliers and to use separate training sets for various weather conditions and/or seasons to enhance the "sensibility" of the model to the weather type.

The results of studying the possibility of filtering the initial data to improve the PVPP generation short-term forecast accuracy allow us to conclude that the application of the k-means methodology is the most effective. The PVPP generation forecast error is reduced by almost 2 times compared with the calculation without filtering and more than 1.5 times compared with the calculation using an empirical filter. The mean absolute percentage error for PVPP day-ahead forecasting with k-means data filtration was calculated to be 18.66%. Thus, it can therefore be concluded that the use of the k-means method for filtering the initial data allows for reducing the PVPP generation forecast error introduced at the stage of forecasting the transparency index to obtain a more accurate forecasting result.

Subsequent improvement of PVPP forecasting accuracy was achieved by adjusting the time-domain of the model by adding the intra-day forecasting procedure, realized in the form of forecasting error prediction. It was assumed that the knowledge of present-day forecasted PVPP generation and the mismatch of the forecasted values compared to the actual data, acquired from irradiance and electrical meters, will give us the opportunity to develop a powerful multi-time-domain forecasting tool. Unlike the other existing approaches, the authors have attempted to establish the bridge between the day-ahead PVPP forecast and the operational hour-ahead PVPP generation forecast by introducing an error forecasting procedure. We provided the comparison of the persistence model, moving average, autoregression, autoregressive moving average and autoregression with exogenous features for short-term forecast error prediction. It was found that the second-order autoregression model AR(2) for short-term forecasting error prediction outperformed all of the methods under consideration.

The developed approach of operational forecasting makes it possible to reduce the total error by almost 1.5 times in comparison with the short-term forecast. The mean average percentage error was calculated to be about 13%. At the same time, the operational forecast makes it possible to obtain more robust estimates of the PVPP generation predicted values: the percentage of operational forecasts meeting the confidence interval of ±1 MW (for 15 MW PVPP) is more than 88%, and the percentage of the forecasts meeting the confidence interval of ±2 MW is more than 97%. Thus, this confirms the effectiveness of the proposed methodology for an operational forecast based on retrospective data on short-term forecasting errors.

The proposed forecasting software was installed on a low-cost distributed computing system, characterized by robust and maintenance-free operation, which is of great importance for power generation facilities operated in the autonomous mode and/or providing system service at the wholesale energy market. Moreover, in the course of PVPP operation, the retrospective dataset is being permanently updated with the newly introduced measurements and calculation results. The proposed hardware system has outstanding scaling properties, so there is no need to introduce high-performance computing facilities even for the Big Data sets.

Author Contributions: Conceptualization, S.A.E. and D.A.S.; data curation, S.A.E., A.I.K., D.A.S. and A.M.R.; formal analysis, V.V.D. and A.M.R.; funding acquisition, A.I.K.; investigation, S.A.E., D.A.S. and D.N.B.; methodology, S.A.E., A.I.K. and D.A.S.; project administration, A.I.K.; resources, S.A.E., V.V.D. and D.N.B.; software, S.A.E., A.I.K., V.V.D. and A.M.R.; Supervision, S.A.E. and A.I.K.; validation, D.A.S., V.V.D., A.M.R. and D.N.B.; visualization, D.A.S.; writing—original draft, S.A.E., A.I.K. and D.A.S.; writing—review and editing, D.N.B. All authors have read and agreed to the published version of the manuscript.

Funding: No funding was received for this study.

Conflicts of Interest: The authors declare no conflict of interest.

References

1. REN21. *Renewables 2017 Global Status Report*; REN21: Paris, France, 2017; ISBN 978-3-9818107-6-9.
2. Stiphout, A.; Brijs, T.; Belmans, R.; Deconinck, G. Quantifying the importance of power system operation constraints in power system planning models: A case study for electricity storage. *J. Energy Storage* **2017**, *13*, 344–358. [CrossRef]
3. Habib, A.; Sou, C.; Hafeez, M.H.; Arshad, A. Evaluation of the effect of high penetration of renewable energy sources (RES) on system frequency regulation using stochastic risk assessment technique (an approach based on improved cumulant). *Renew. Energy* **2018**, *127*, 204–212. [CrossRef]
4. Tielens, P.; Van Hertem, D. The relevance of inertia in power systems. *Renew. Sustain. Energy Rev.* **2016**, *55*, 999–1009. [CrossRef]
5. Farhoodnea, M.; Mohamed, A.; Shareef, H.; Zayandehroodi, H. Power quality impact of renewable energy-based generators and electric vehicles on distribution systems. *Procedia Technol.* **2013**, *11*, 11–17. [CrossRef]
6. Triviño-Cabrera, A.; Longo, M.; Foiadelli, F. Impact of renewable energy sources in the power quality of the Italian electric grid. In Proceedings of the 11th IEEE International Conference on Compatibility, Power Electronics and Power Engineering, Cadiz, Spain, 4–6 April 2017; pp. 576–581.
7. Balaban, G.; Lazaroiu, G.C.; Dumbrava, V.; Sima, C.A. Analysing Renewable Energy Source Impacts on Power System National Network Code. *Inventions* **2017**, *2*, 23. [CrossRef]

8. Lee Hau Aik, D.; Andersson, G. Impact of Renewable Energy Sources on Steady-state Stability of Weak AC/DC Systems. *CSEE J. Power Energy Syst.* **2017**, *3*, 319–430. [CrossRef]
9. Ameur, A.; Loudiyi, K.; Aggour, M. Steady State and Dynamic Analysis of Renewable Energy Integration into the Grid using PSS/E Software. *Energy Procedia* **2017**, *141*, 119–125. [CrossRef]
10. Tonkoski, R.; Turcotte, D.; El-Fouly, T.H.M. Impact of High PV Penetration on Voltage Profiles in Residential Neighborhoods. *IEEE Trans. Sustain. Energy* **2012**, *3*, 518–527. [CrossRef]
11. Petinrin, J.O.; Shaaban, M. Impact of renewable generation on voltage control in distribution systems. *Renew. Sustain. Energy Rev.* **2016**, *65*, 770–783. [CrossRef]
12. Begovic, M.; Pregelj, A.; Rohatgi, A.; Novosel, D. Impact of renewable distributed generation on power systems. In Proceedings of the 34th Annual Hawaii International Conference on System Science, Maui, HI, USA, 3–6 January 2001; pp. 654–663.
13. Essallah, S.; Bouallegue, A.; Khedher, A. Optimal Sizing and Placement of DG Units in Radial Distribution System. *Int. J. Renew. Energy Res.* **2018**, *8*, 166–167.
14. Lajda, P. Short-Term Operation Planning in Electric Power Systems. *J. Oper. Res. Soc.* **1981**, *32*, 675–682. [CrossRef]
15. Navarro, R. Short and medium term operation planning in electric power systems. *IEEE/PES Power Syst. Conf. Expo.* **2009**, *1*, 1–8.
16. Talari, S.; Shafie-khah, M.; Osório, G.J.; Aghaei, J.; Catalão, J.P. Stochastic modelling of renewable energy sources from operators' point of-view: A survey. *Renew. Sustain. Energy Rev.* **2018**, *81*, 1953–1965. [CrossRef]
17. Dai, H.; Zhang, N.; Su, W. A Literature Review of Stochastic Programming and Unit Commitment. *J. Power Energy Eng.* **2015**, *3*, 206–214. [CrossRef]
18. Aien, M.; Rashidinejad, M.; Firuz-Abad, M.F. Probabilistic power flow of correlated hybrid wind-PV power systems. *IET Renew. Power Gener.* **2014**, *8*, 649–658. [CrossRef]
19. Zachary, S.; Dent, C.J. Probability theory of capacity value of additional generation. *J. Risk Reliab.* **2012**, *226*, 33–43. [CrossRef]
20. Ioannou, A.; Angus, A.; Brennan, F. Risk-based methods for sustainable energy system planning: A review. *Renew. Sustain. Energy Rev.* **2017**, *74*, 602–615. [CrossRef]
21. Ayodele, T.R.; Ogunjuyigbe, A.S.O.; Akpeji, K.O.; Akinola, O.O. Prioritized Rule Based Load Management Technique for Residential Building Powered by PV/Battery System. *Eng. Sci. Technol. Int. J.* **2017**, *20*, 859–873. [CrossRef]
22. Wan, C.; Zhao, J.; Song, Y.; Xu, Z. Photovoltaic and solar power forecasting for smart grid energy management. *CSEE J. Power Energy Syst.* **2015**, *1*, 38–46. [CrossRef]
23. Prema, V.; Rao, K.U. Development of statistical time series models for solar power prediction. *Renew. Energy* **2015**, *83*, 100–109. [CrossRef]
24. Kaplanis, S.; Kaplani, E. A model to predict expected mean and stochastic hourly global solar radiation I(h;nj) values. *Renew. Energy* **2007**, *32*, 1414–1425. [CrossRef]
25. Ferrari, S.; Lazzaroni, M.; Piuri, V.; Cristaldi, L.; Faifer, M. Statistical models approach for solar radiation prediction. *IEEE Int. Instrum. Meas. Technol. Conf.* **2013**, *1*, 1734–1739.
26. Li, Y.; He, Y.; Su, Y.; Shu, L. Forecasting the daily power output of a grid-connected photovoltaic system based on multivariate adaptive regression splines. *Appl. Energy* **2016**, *180*, 392–401. [CrossRef]
27. Perez, R.; Ineichen, P.; Moore, K.; Kmiecik, M.; Chain, C.; George, R.; Vignola, F. A new operational model for satellite-derived irradiances: Description and validation. *Sol. Energy* **2002**, *73*, 307–317. [CrossRef]
28. Mathiesen, P.; Collier, C.; Kleissl, J. A high-resolution, cloud-assimilating numerical weather prediction model for solar irradiance forecasting. *Sol. Energy* **2013**, *92*, 47–61. [CrossRef]
29. Gohari, M.I.; Urquhart, B.; Yang, H.; Kurtz, B.; Nguyen, D.; Chow, C.W.; Ghonima, M.; Kleissl, J. Comparison of solar power output forecasting performance of the total sky imager and the University of California, San Diego Sky Imager. *Energy Procedia* **2014**, *49*, 2340–2350. [CrossRef]
30. Larson, D.P.; Nonnenmacher, L.; Coimbra, C.F.M. Day-ahead forecasting of solar power output from photovoltaic plants in the American Southwest. *Renew. Energy* **2016**, *91*, 11–20. [CrossRef]
31. Chaouachi, A.; Kamel, R.M.; Ichikawa, R.; Hayashi, H.; Nagasaka, K. Neural Network Ensemble-based Solar Power Generation Short-Term Forecasting. *Int. J. Inf. Math. Sci.* **2009**, *5*, 332–337. [CrossRef]

32. Monteiro, C.; Santos, T.; Fernandez-Jimenez, L.A.; Ramirez-Rosado, I.J.; Terreros-Olarte, M.S. Short-term power forecasting model for photovoltaic plants based on historical similarity. *Energies* **2013**, *6*, 2624–2643. [CrossRef]
33. Zeng, J.; Qiao, W. Short-term solar power prediction using a support vector machine. *Renew. Energy* **2013**, *52*, 118–127. [CrossRef]
34. Persson, C.; Bacher, P.; Shiga, T.; Madsen, H. Multi-site solar power forecasting using gradient boosted regression trees. *Sol. Energy* **2017**, *150*, 423–436. [CrossRef]
35. Voyant, C.; Muselli, M.; Paoli, C.; Nivet, M. Numerical weather prediction (NWP) and hybrid ARMA/ANN model to predict global radiation. *Energy* **2012**, *39*, 341–355. [CrossRef]
36. Zheng, F.; Zhong, S. Time series forecasting using a hybrid RBF neural network and AR model based on binomial smoothing. *Int. J. Math. Comput. Sci.* **2011**, *75*, 1471–1475.
37. Khan, I.; Zhu, H.; Yao, J.; Khan, D.; Iqbal, T. Hybrid Power Forecasting Model for Photovoltaic Plants Based on Neural Network with Air Quality Index. *Int. J. Photoenergy* **2017**, *1*. [CrossRef]
38. Wang, J.; Jiang, H.; Wu, Y.; Dong, Y. Forecasting solar radiation using an optimized hybrid model by Cuckoo Search algorithm. *Energy* **2015**, *81*, 627–644. [CrossRef]
39. Snegirev, D.A.; Eroshenko, S.A.; Valiev, R.T.; Khalyasmaa, A.I. Algorithmic Realization of Short-term Solar Power Plant Output Forecasting. In Proceedings of the II International Conference on Control in Technical Systems (CTS'2017), Saint Petersburg, Russia, 25–27 October 2017; pp. 49–52.
40. Snegirev, D.A.; Valiev, R.T.; Eroshenko, S.A.; Khalyasmaa, A.I. Functional assessment system of solar power plant energy production. In Proceedings of the International Conference on Energy and Environment (CIEM), Bucharest, Romania, 19–20 October 2017; pp. 349–353.
41. Matuszko, D. Influence of the extent and genera of cloud cover on solar radiation intensity. *Int. J. Climatol.* **2012**, *32*, 2403–2414. [CrossRef]
42. Kanungo, T.; Mount, D.; Netanyahu, N.; Piatko, C.; Silverman, R.; Wu, A. An efficient k-means clustering algorithm: Analysis and implementation. *IEEE Trans. Pattern Anal. Mach. Intell.* **2000**, *24*, 881–892. [CrossRef]
43. Boland, J. Time series and statistical modelling of solar radiation. In *Recent Advances in Solar Radiation Modelling*; Springer: Berlin/Heidelberg, Germany, 2008; pp. 283–312.
44. Romanov, A.M.; Romanov, M.P.; Manko, S.V.; Volkova, M.A.; Chiu, W.-Y.; Ma, H.-P.; Chiu, K.-Y. Modular Reconfigurable Robot Distributed Computing System for Tracking Multiple Objects. *IEEE Syst. J.* **2020**. [CrossRef]
45. Romanov, A.M. A novel architecture for Field-Programmable Gate Array-based Ethernet Powerlink controlled nodes. *Tr. MAI* **2019**, *106*, 15–30.
46. Romanov, A.M.; Slaschov, B.V. FPGA-based Kalman filtering for motor control. In *Network Security and Communication Engineering*; CRC Press: Boca Raton, FL, USA, 2015; pp. 569–572. [CrossRef]
47. Romanov, A.M. A review on control systems hardware and software for robots of various scale and purpose. Part 3. Extreme robotics. *Russ. Technol. J.* **2020**, *8*, 14–32. (In Russian) [CrossRef]
48. Romanov, A.; Bogdan, S. Open source tools for model-based FPGA design. In Proceedings of the 2015 International Siberian Conference on Control and Communications (SIBCON), Omsk, Russia, 21–23 May 2015; pp. 1–6. [CrossRef]

Article

Hybridizing Deep Learning and Neuroevolution: Application to the Spanish Short-Term Electric Energy Consumption Forecasting

Federico Divina [1,2,*,†], José F. Torres [1,†], Miguel García-Torres [1,2], Francisco Martínez-Álvarez [1] and Alicia Troncoso [1]

1 Data Science and Big Data Lab, Pablo de Olavide University, ES-41013 Seville, Spain; jftormal@upo.es (J.F.T.); mgarciat@upo.es (M.G.-T.); fmaralv@upo.es (F.M.-Á.); atrolor@upo.es (A.T.)
2 Computer Engineer Department, Universidad Americana de Paraguay, Asunción 1029, Paraguay
* Correspondence: fdivina@upo.es
† These authors contributed equally to this work.

Received: 1 July 2020; Accepted: 5 August 2020; Published: 7 August 2020

Abstract: The electric energy production would be much more efficient if accurate estimations of the future demand were available, since these would allow allocating only the resources needed for the production of the right amount of energy required. With this motivation in mind, we propose a strategy, based on neuroevolution, that can be used to this aim. Our proposal uses a genetic algorithm in order to find a sub-optimal set of hyper-parameters for configuring a deep neural network, which can then be used for obtaining the forecasting. Such a strategy is justified by the observation that the performances achieved by deep neural networks are strongly dependent on the right setting of the hyper-parameters, and genetic algorithms have shown excellent search capabilities in huge search spaces. Moreover, we base our proposal on a distributed computing platform, which allows its use on a large time-series. In order to assess the performances of our approach, we have applied it to a large dataset, related to the electric energy consumption registered in Spain over almost 10 years. Experimental results confirm the validity of our proposal since it outperforms all other forecasting techniques to which it has been compared.

Keywords: time-series forecasting; deep learning; evolutionary computation; neuroevolution

1. Introduction

The electric energy needs are constantly growing. It is estimated that such demand will increment from 549 quadrillion British thermal unit (Btu), registered in 2012, to 629 quadrillion Btu in 2020. A further increment of 48% is estimated by 2040 [1].

The accurate estimation of the short-term electric energy demand provides several benefits. The economic benefits are evident because this would allow us to allocate only the right amount of resources that are needed in order to produce the amount of energy actually needed to face the actual demand [2,3]. There are also environmental aspects to consider, since, by producing only the right amount of energy required, the emission of CO_2 would be reduced as well. In fact, energy efficiency is another relevant goal pursued with these kinds of approaches since the accurate forecasting of electricity demand in public buildings or in industrial plants usually leads to energy savings [4–6].

Such observations highlight the importance of being able to count on efficient electric energy management systems and prediction strategies and, consequently, different organizations around the world are taking actions in order to increase energy efficiency. Hence, the European Union (EU), under the current energy plan [7], established that EU countries will have to embrace various energy efficiency requirements with the objective of improving at least a 20% the energy efficiency. In addition to this, countries belonging to the EU closed an agreement to obtain an additional 27% increment of the efficiency by 2020, with the possibility of increasing the target to 30% by the year 2030.

Forecasting algorithms could contribute to reaching such objectives [2,3]. In this context, energy demand forecasting can be described as the problem of predicting the energy demand within a specified prediction horizon, using past data, or, in other words, a historical window.

Depending on the time scale of the predictions, we can generally distinguish three classes of forecasting, i.e., short, medium and long-term forecasting. In short-term forecasting, the objective is to predict the energy demand using horizons going from one hour up to a week. If the prediction horizon is set between one week and one month, we talk about medium-term forecasting, while long-term forecasting involves longer horizons [8].

In this paper, we focus on the problem of short-term forecasting. This is an important problem, since with accurate predictions of short-term load it would be possible to make precisely plan the resources that need to be allocated in order to face the actual demand, which, as already stated, would have benefits from both the economical and environmental points of view.

To this aim, we propose an extension of the work proposed in [9], where a deep feed-forward neural network was used to tackle the short-term load forecasting problem. In the original work, the tools provided by the H2O big data analysis framework were used along with the Apache Spark platform for distributed computing.

Differently from [9], where a grid search strategy was used for setting the values of the deep neural network parameters, in this work, we propose to use a genetic algorithm (GA) in order to determine a sub-optimal set of hyper-parameters for building the deep neural network that will then be used for obtaining the predictions. Due to the large search space composed of all hyper-parameters of a deep learning network, and considering that the method should be scalable for big data environments, it has been decided to reduce the search range of the GA. For this reason, our proposal will not always be able to find the optimal set of hyper-parameters for the network, but ensures a competitive sub-optimal configuration.

Our main motivation lies in the observation that the success of deep learning depends on finding an architecture to fit the task. As deep learning has scaled up to more challenging problems, the architectures have become difficult to design by hand [10]. To this aim, evolutionary algorithms (EAs) can be used in order to find good configurations of the deep neural networks. Individuals can be set of parameter values, and their fitnesses are determined based on how well they can be trained to perform in the task.

This field is known as neuroevolution, which, in a nutshell, can be defined as a strategy for evolving neural networks with the use of EAs [11]. Usually, deep artificial neural networks (DNNs) are trained via gradient-based learning algorithms, namely backpropagation, see for example [12]. EAs can be used in order to seek the optimal values of hyper parameters, for the example the learning rates, or the number of layers and the amount of neurons per layer, among others.

It has been proven that EAs can be combined with backpropagation-based techniques, such as Q-learning and policy gradients, on difficult problems, see, e.g., [13]. In fact, the problem of setting parameters for such methods is not trivial, and, if the parameters are not correctly set, the forecasting can be poor.

The above observations motivate us to use a neuroevoltution approach in order to tackle the short-term energy load forecasting problem. In order to validate our proposal, we applied it to a dataset regarding the electric energy consumption registered over almost 10 years in Spain. We have also compared our proposal with other standard and machine learning (ML) strategies, and results obtained confirm that our proposal achieves the best predictions.

In the following, we summarise the main contributions of this paper:

1. We propose a new general-purpose approach based on deep learning for big data time-series forecasting. Due to the high computational cost of the deep learning, we adopted a distributed computing solution in order to be able to process large time series.
2. The hyper-parameter tuning and optimization of the deep neural networks is a key factor for obtaining competitive results. Usually, the hyper-parameters of a deep neural network are pre-fixed previously or computed by a grid search, which performs an exhaustive search through the whole set of established hyper-parameters. However, the grid search presents an important limitation: it works with discrete values, which greatly limits the fine-tuning of the vast majority of hyper-parameters. Thus, an evolutionary search is proposed to find the hyper-parameters.
3. We conduct a wide experimentation using Spanish electricity consumption registered over 10 years, with measurements recorded every 10 min. Results show a mean relative error of 1.44%, demonstrating the high potential of the proposed approach, also compared to other forecasting strategies.
4. We evaluate our proposal predictive accuracy and compare it with a strategy based on deep learning using a grid search for setting the hyper parameters. The evolutionary search showed to be effective in order to achieve higher accuracy.
5. In addition, we compare the approach with seven state-of-the-art forecasting algorithms such as ARIMA, decision tree, an algorithm based on gradient boosting, random forest, evolutionary decision trees, a standard neural network and an ensemble proposed in [14], outperforming all of them.
6. We analyze how the size of the historical window affects the accuracy of the model. We found that when using the past 168 values as input features to predict the next 24 values the best results were obtained.

The rest of the paper is organized as follows. In Section 2 we provide a brief overview of the state of the art of electric energy time-series forecasting. The dataset used in this work is described and analyzed in Section 3.1, while the methodology used is discussed in Section 3.2. In Section 4 we describe the results obtained by our approach and compare them to those achieved by other strategies. Finally, we draw the main conclusions and identify futures works in Section 5.

2. Related Works

As previously mentioned, a lot of attention has been paid to short-term electricity consumption forecasting during the last decades. This section provides a brief overview of up-to-date related works.

We can distinguish two main strategies to predict energy consumption. A first strategy is based on conventional methods, e.g., [15,16], whilst an alternative, and more recent strategy, is based on ML techniques.

Conventional methods include, among others, statistical analysis, smoothing techniques such as the autoregressive integrated moving average (ARIMA), exponential smoothing and regression-based approaches. Such techniques can obtain satisfactory results when applied to linear problems.

In contrast, ML strategies are also suitable for non-linear cases. We refer the reader to [17] for an expanded survey on data mining techniques applied to electricity-related time-series forecasting. In this work, several markets and prediction horizons are considered and discussed.

Popular ML techniques successfully applied to the forecasting of power consumption data include Artificial Neural Networks (ANN) [18–20] or Support Vector Machines (SVM), see, for instance, [21,22].

Other strategies are based on pattern similarity [23,24]. Since 2011, when the Pattern Sequence based Forecasting (PSF) algorithm was published [24], a number of variants has been proposed for forecasting this kind of time-series [25–28], including an R package [29] and a big data version [30]. Grey forecast models have also been used for predicting time-series. In particular such an approach has been applied to forecast the demand of natural gas in China. For instance, in [31] a self-adapting intelligent grey prediction model was proposed, where a linear function was used in order to automatically optimize the parameters used by the proposed grey model. This strategy was substituted with a genetic algorithm in [32], which resolved various limitations of the previous mechanism. A novel time-delayed polynomial grey model was introduced in [33], while in [34] authors proposed a least squares support vector machine model based on grey analysis.

Recently, Deep Learning (DL) has also been applied to this problem, see, e.g., [9,35]. However, to the best of our knowledge, a part from the early version [36] and few other works, such as [37], in which Brazilian data were analyzed, or [38] for Irish data, or [39] for Chinese data, no other works based on DL can be found in the literature.

Although ML techniques provide effective solutions for time-series forecasting, these methods tend to get stuck in a local optimum. For instance, ANN and SVM may get trapped in a local optimum if their configuration parameters are not properly set.

Recently, methods developed for big data environments have also been applied to electricity consumption forecasting. In [40] an approach based on the *k*-weighted nearest neighbours algorithm was introduced and implemented using the Apache Spark framework. The performances of the resulting algorithm were tested using a Spanish energy consumption Big Data time-series. As mentioned above, in 2018, Torres et al. [9] proposed a DL model to deal with big data time-series forecasting. In particular, the H2O Big Data analysis framework was used. Results from a real-world dataset composed of electricity consumption in Spain, with a ten-minute frequency sampling rate, from 2007 to 2016 were reported.

As can be seen, although much attention has been paid to the electricity consumption forecasting problem, few works based on DL have been proposed. Moreover, such existing works did not applied any metaheuristic strategy to set the parameters. These facts highlight the existing gap in the literature and justify, from the authors' point of view, the development of this work.

As previously stated, in this paper we aim at using DL, in order to perform time-series forecasting. In DL, many parameters have to be set. The setting of such parameters have a great influence on the final results obtained by such a strategy. An alternative way to set the DL parameters is to use an Evolutionary Algorithm (EA) in order to find a sub-optimal set of parameters. This field, known as neuroevolution [11,41], has received much attention lately in the ML community. Neuroevolution enables important capabilities such as learning neural network building blocks, e.g., the activation function, hyperparameters, architectures and even the algorithms for learning themselves. Neuroevolution also differs from DL (and deep reinforcement learning) since in neuroevolution a population of solutions is maintained during the search. This provides extreme exploration capabilities and the possibility of massive parallelization. There also exist alternative strategies in order to find an optimal set of parameter, going from grid search to more complex approaches, such as methods based on Bayesian optimization, see, for instance [42,43]. Neuroevolution has been successfully applied to different fields, especially in image classification, where Convolutional Neural Networks (CNN) are evolved, see, for instance [44–47]. To be best of our knowledge, Neuroevolution has not been applied to time-series forecasting.

3. Data and Methodology

3.1. Data

In order to assess the quality of our proposal, we used a dataset containing information regarding the global electricity consumption registered in Spain (in MW), available at [48].

In particular, the data were recorder over a period going from 1 January 2007 at midnight until 21 June 2016 at 11:40 pm, which amounts to nine years and six months. Specifically, the data is relative to the consumption measured at 10 minutes intervals, meaning that the dataset consists of a total of 497,832 measurements. No missing values or outliers were found, since data are provided by the Spanish Nominated Electricity Market Operator (NEMO) and all data are already preprocessed and cleaned.

Time-series regarding the electric energy demand are typically non-stationary. This fact renders the problem of forecasting the electric energy demand challenging, since such time-series present statistical properties, such as the mean, variance and autocorrelation, that are not all constant over time. It follows that they can present changes in variance, trends or seasonal effects. For this reason, we performed a preliminary study of the dataset in order to assess whether or not the time-series used in this paper is stationary. To this aim, we analyzed the AutoCorrelation Function (ACF) and the Partial AutoCorrelation Function (PACF) of the time-series, which are reported in Figure 1.

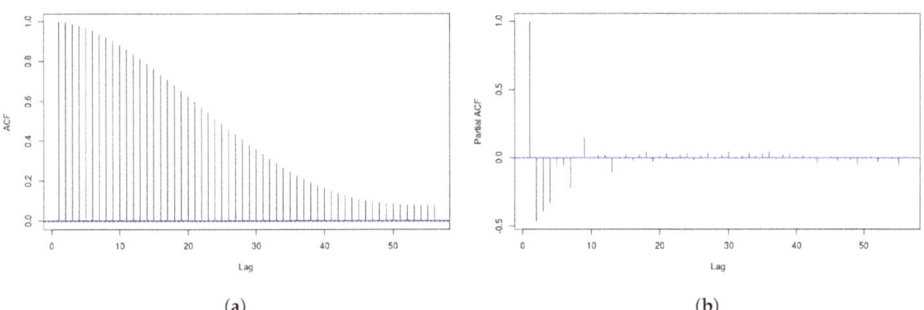

Figure 1. Correlation plots for the original time-series. (**a**) AutoCorrelation Function (ACF); (**b**) Partial AutoCorrelation Function (PACF).

From Figure 1a, we can notice that the time-series has a high correlation with a number significant of lags, while from Figure 1b we can see that there are four spikes in the first lags, from which we can determine the order of autoregression of the time-series. From these observations, we can conclude that the time-series is not stationary, and that the order of autoregression to be used should be 4.

A preprocessing of the dataset had to be applied before it could be used. In particular, we used the preprocessing strategy proposed in [36], which is graphically depicted in Figure 2. In a first step, we extract the attribute corresponding to the energy consumption, obtaining in this way a consumption vector V_c.

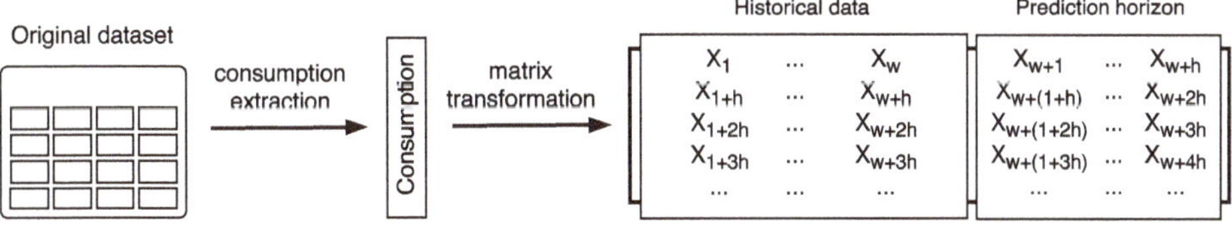

Figure 2. Dataset pre-processing. w determines the amount of historical data used, while h represents the prediction horizon.

From V_c matrix M_c is built. The size of M_c depends on the values of the historical window (w) and of the prediction horizon (h) used. Notice that w determines the number of previous entries that will be used in order to induce a forecasting model that will be used to estimate the subsequent h values.

In this work, as in [36], h was set to 4 hours, which corresponds to a value of 24 reads. Various values of w were tested.

In particular, w was set to values 24, 48, 72, 96, 120, 144 and 168. Such values correspond to 4, 8, 12, 16, 20, 24 and 28 hours, respectively.

One the matrix M_c has been obtained, we divided the resulting dataset into a 70% used as a training set, while the remaining 30% was used as a testing set. This means that the prediction model was obtained using only the training set. The forecasting performances of the so induced model are assessed on the test set, which basically represents unseen data. Within the training set, a 30% is used as a validation set for determining the deep learning hyperparameters.

These preprocessing steps yield the generation of seven different matrices, whose information is reported in Table 1. Note that for all the obtained datasets, the last 24 columns represent the prediction horizon.

Table 1. Dataset information depending on the value of w.

w	#Rows	#Columns	File Size (In MB)
24	20,742	48	6
48	20,741	72	9
72	20,740	96	11.9
96	20,739	120	14.9
120	20,738	144	17.9
144	20,737	168	20.9
168	20,736	192	23.9

3.2. Methodology

This section describes the proposed methodology for forecasting time-series using a deep learning approach. There are various deep learning architectures which can be used for time-series forecast, such as convolutional neural nets (CNN), recurrent neural nets (RNN) or feed-forward neural nets (FFNN).

In this paper, a deep feed-forward network has been used, implemented by R package H2O [49]. H2O is an open-source framework that implements various machine learning techniques in a parallel and distributed way using a single machine or a cluster of machines, being scalable for big data projects.

Among the algorithms included in H2O, we can find a feed-forward neural network, that is the most common network architectures. The main characteristic of this net is that each neuron is a basic element of processing and their information is propagated through adjacent neurons.

In addition, in order to select the configuration of the network hyperparameters, we used a GA, which was implemented by using the GA R package [50].

3.2.1. Parameters of the Neural Network

The network architecture implemented in the H2O package needs to be configured by setting different parameters, that will affect the behavior of the neural network and influence the final results. The most important parameters are: number of layers, neurons per hidden layer, L1 (λ), ρ, ϵ, activation and distribution functions and end metric. These are the parameters that the GA will optimize.

The parameter λ controls the regularization of the model by inserting penalties in the model creation process in order to adjust the predictions as much as possible with actual values and the penalization is defined by the following equation:

$$\lambda \sum_{i=0}^{n} |w_i| \,. \tag{1}$$

In Equation (1), n is the number of weights received by the neurons and w_i represents the weight for the neuron i.

The parameter ρ allows us to manage the update of different weights of synapses and is used to maintain some consistency between the different updates of previous weights.

The parameter ϵ prevents the deep learning algorithm from being stuck in local optimums or to skip a global optimum, and can assume values between 0 and 1.

The activation function can assume three values: tanh (hyperbolic tangent), ramp function, maxout.

Seven different possibilities are considered for the distribution function: Gaussian, Poisson, Laplace, Tweedie, Huber, Gamma and Quantile.

The end metric defines the specific measure that is used to stop early the training phase of the deep learning algorithm. There are seven different possibilities: mean squared error (MSE), Deviance (the difference between an expected value and an observed value), root mean squared error (RMSE), mean absolute error (MAE), root mean squared log error (RMSLE), the mean per class error and lift top group. The last metric is a measure of the relative performance.

The possible values for each parameter are shown in Table 2.

Table 2. Search space of the neural network parameters.

Parameter	Values
Layers	From 2 to 100
Neurons	From 10 to 1000
Lambda (λ)	From 0 to 1×10^{-10}
Rho (ρ)	From 0.99 to 1
Epsilon (ϵ)	From 0 to 1×10^{-12}
Activation function	From 0 to 3
Distribution function	From 0 to 7
End metric	From 0 to 7

As we described before, the activation function, distribution function and end metric are categorical parameters, so each value corresponds to a specific category of the parameter.

3.2.2. Genetic Algorithm Parameters

As previously stated, in order to find a sub-optimal set of hyper-parameters, described in the previous section, for the deep learning algorithm, we use a GA. In particular we use the implementation provided by the GA R package [50]. So our proposal lies within the field of neuroevolution.

The GA package contains a collection of general-purpose functions for optimization using genetic algorithms. The package includes a flexible set of tools for implementing genetic algorithms in both the continuous and discrete case, whether constrained or not. However the package does not allow to simultaneously optimize continuous and discrete parameters, so we had to treat all the parameters as continuous, which caused the dimension of the search space to increase drastically.

The package allows us to define objective functions to be optimized, which, in our case, is the forecasting results obtained by a deep neural network built with a specific set of parameters. In fact, each individual of the population encodes the values of the eight parameters shown in Table 2.

Each parameter setting yields a specific deep neural network, which is then applied to the data and the forecasting result represent the fitness of the individual.

In particular, the fitness of an individual is equal to the MRE obtained by the deep neural network on the validation set, being the MRE defined as:

$$MRE = \frac{1}{n} \sum_{i=1}^{n} \frac{|Y_i - \hat{Y}_i|}{Y_i}, \tag{2}$$

where $\hat{Y}_i$ is the predicted value, Y_i the real value and $\overline{Y}_i$ is the mean of the observed data, and n is the number of data.

Several genetic operators are available and can be combined to explore the best settings for the current task. After having performed a set of preliminary experiments aimed at setting the GA's parameters, we used, in our implementation, a tournament selection mechanism (with tournament size of 3), the BLX-a crossover (with $a = 0.5$), which combines two parents to generate offspring by sampling a new value in a defined range with the maximum and the minimum of the parents [51]. We used the random mutation around the solution, which allows us to change one value of an element by another value.

The setting of the parameters used in the GA are reported in Table 3. The value shown are those that obtained the best performances in the preliminary runs, but the population size. In fact, better results were achieved with higher population size. However, the computational cost increases dramatically the higher the population size is. In fact, the deep learning algorithm takes around 89.42 s for a number of layers between 2 and 100 and for a number of neurons between 10 and 1000.

The execution of the GA with the deep learning algorithm as a fitness function and with the parameters defined in Table 3 takes around five days. If the population size is doubled, the execution can take more than one week. It is necessary to enhance one of the parameters (population size or number of generations) but not both. Moreover, if the fitness of the best individual does not improve after 50 generations, the GA is stopped.

At the end of the execution, the best individual is returned and used in order to build a deep learning network.

Table 3. Genetic algorithm (GA) parameter setting.

Operator	Value
Population size	50
Generations	100
Limit of generations	50
Crossover probability	0.8
Mutation probability	0.1
Elitisms probability	0.05

3.2.3. Description of the Methodology

The main objective of this work is to predict the next h future values, called the prediction horizon, of a time-series $[x_1, x_2, \ldots, x_t]$.

The predictions are based on w previous values, or, in other words, on a historical data window. This process is called multi-step forecasting, as various consecutive values have to be predicted. The aim of multi-step forecasting is to induce a prediction model f, and in our case f is obtained by using a deep learning strategy, following the equation:

$$[x_{t+1}, x_{t+2}, \ldots, x_{t+h}] = f(x_t, x_{t-1}, \ldots, x_{t-(w-1)}). \tag{3}$$

Unfortunately, frameworks that provide deep learning networks model, such as H20, does not support this multi-step formulation.

In order to solve this issue, a different methodology has been proposed [9]. The basic idea is to divide the main problem into h prediction sub-problems. Then a forecasting model will be induced for each of the sub-problems, as shown in Equation (4).

$$\begin{aligned} x_{t+1} &= f_1(x_t, x_{t-1}, \ldots, x_{t-(w-1)}) \\ x_{t+2} &= f_2(x_t, x_{t-1}, \ldots, x_{t-(w-1)}) \\ \ldots &= \ldots \\ x_{t+(h-1)} &= f_{(h-1)}(x_t, x_{t-1}, \ldots, x_{t-(w-1)}) \\ x_{t+h} &= f_h(x_t, x_{t-1}, \ldots, x_{t-(w-1)}) \end{aligned} \tag{4}$$

Notice that in this way, we lose the time relationship between consecutive records of the time-series. For instance, instants $t+1$, $t+2$, $t+3$ or $t+4$ will not be considered when forecasting $t+5$.

On the other hand, considering such values for the predictions could increment the forecasting error. This is because values for $t+1$, $t+2$, $t+3$ or $t+4$ are based on predictions, and they would have a negative effect on the forecasts if the values were not precisely estimated.

It follows that a search for optimal parameters should be carried out for each sub-problem, where the evaluation of each individual corresponds to the error made by the neural network in the training phase. This means that the computational time needed to train the complete model is high. However, the capability of H2O to perform distributed computation decreases the total computational time required.

4. Experimental Results

In this section, we present the forecast results obtained on the dataset described in Section 3.1 by the strategy we propose. We also present a comparison with different methods, both standard and ML based.

In order to assess the predictions produced by our proposal, we used the *MRE* measure, as defined in Equation (2). *MRE* represents the ratio of the forecasting absolute error to the observed value.

Before presenting the comparison with other methods, we inspect the results obtained by the proposed strategy for each historical window value used (w) and each subproblem (h). Figure 3 shows a graphical representation of the results obtained, showing the associated MRE for different values of w, when varying the length of h. We can see that the best results were achieved when the forecasting is based on more historical data, i.e., for higher values of w. In fact, the best results were obtained for $w = 168$. Analogously, the MRE increases as h becomes longer. The proposed strategy obtains similar results for $w = \{168, 144, 120\}$ on all the considered values of the prediction horizon h.

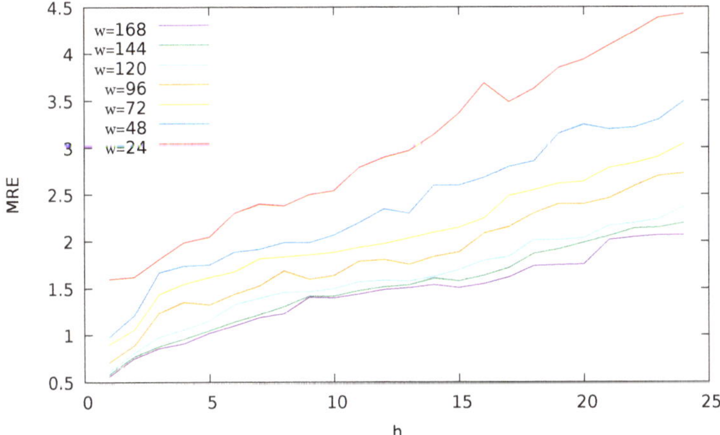

Figure 3. Results obtained for each value of h and w.

It can be noticed that there is a significant increment in the error when the historical window size is lower. In particular, when w is set to 24 or 48, the predictions degenerates evidently. We can also notice that performances of the proposed strategy deteriorates, i.e., the achieved MRE is higher, as the values of h increase. This means that it is more difficult to predict further in the future.

Table 4 shows the parameters selected by the GA for each h when a historical window of 168 was used. We can notice that the number of layers range between 27 and 98, and the number of neurons per layer between 478 and 942. It does not seem that this parameter is connected with the value of h.

Parameters λ, ρ and ϵ assume almost the same values on all the cases, while the *Maxout* is the activation function mostly chosen. The GA selected two possibilities as distribution functions, namely the *Gaussian* and the *Huber* function. The end metric selected, on the other hand, presents more variations. This could suggest that we could perhaps fix some of the parameters, e.g., ϵ, in order to reduce the search space.

Table 4. Parameters found by the GA for $w = 168$.

h	Layers	Neurons	λ	ρ	ϵ	Activation	Distribution	End Metric
1	52	942	4.09×10^{-10}	1.00	6.43×10^{-12}	Tanh	Gaussian	Deviance
2	68	921	0	1.00	0	Maxout	Huber	MSE
3	75	880	0	1.00	0	Maxout	Huber	Deviance
4	68	921	0	1.00	0	Maxout	Huber	MSE
5	88	504	0	1.00	0	Maxout	Huber	Deviance
6	80	789	0	1.00	0	Maxout	Huber	MSE
7	74	892	0	1.00	0	Maxout	Huber	RMSLE
8	46	300	0	1.00	0	Maxout	Huber	MAE
9	75	889	5.57×10^{-10}	0.99	6.74×10^{-10}	Tanh	Gaussian	Mean per class error
10	25	852	0	1.00	0	Maxout	Huber	RMSLE
11	58	843	3.69×10^{-10}	1.00	2.45×10^{-10}	Tanh	Gaussian	RMSE
12	41	491	0	1.00	0	Maxout	Huber	RMSLE
13	17	552	0	0.99	0	Maxout	Huber	MSE
14	26	661	0	0.99	0	Maxout	Huber	MAE
15	89	811	5.61×10^{-10}	0.99	4.23×10^{-10}	Tanh	Gaussian	RMSE
16	98	697	0	1.00	0	Maxout	Huber	MAE
17	74	478	1.46×10^{-10}	1.00	3.58×10^{-10}	Tanh	Gaussian	Deviance
18	62	705	2.74×10^{-10}	0.99	6.64×10^{-10}	Tanh	Gaussian	MAE
19	65	879	0	0.99	0	Maxout	Huber	MAE
20	81	780	7.62×10^{-10}	0.99	5.21×10^{-10}	Tanh	Gaussian	MSE
21	27	931	0	1.00	0	Maxout	Huber	MAE
22	95	745	0	1.00	0	Maxout	Huber	Deviance
23	41	923	0	1.00	0	Maxout	Huber	MSE
24	80	754	0	1.00	0	Maxout	Huber	MAE

As previously stated, in order to globally assess the performance of our proposal, we compared the results achieved by our methodology (NDL) with the results obtained by other strategies commonly used for time-series forecast. In particular, we considered Random Forest (RF), Artificial Neural Networks (NN), Evolutionary Decision Trees (EV), the Auto-Regressive Integrated Moving Average (ARIMA), an algorithm based on Gradient Boosting (GBM), three Deep Learning models (FFNN, Feed-Forward Neural Network; CNN, Convolutional Neural Network; LSTM, Long Short-Term Memory), decision tree algorithm (DT) and an ensemble strategy that was proposed in [14], which combined regression trees-based, artificial neural networks and random forests (ENSEMBLE).

For ARIMA, we used the tool in Ref. [52] for determining the order of auto-regressive (AR) terms (p), the degree of differencing (d) and the order of moving-average (MA) terms (q). The values obtained are $p = 4$, $d = 1$ and $q = 3$. The value for the auto-regressive parameter and the degree of differencing confirm that the time-series is not stationary, as indicated in Section 3.1.

The deep learning models were designed using *H2O* framework of R [49]. The difference between NDL and DL, is that in the latter case, the network is trained with stochastic gradient descend using back-propagation algorithm. In order to set the parameters for DL, we used a grid search approach. As a consequence, we used a hyperbolic tangent function as activation function, the number of hidden layer was set to 3 and the number of neurons to 30. The distribution function was set to Poisson and in order to avoid overfitting, the regularization parameter (Lambda) has been set to 0.001. The other two parameters (ρ and ϵ) were set as default as in [36].

The DT algorithm is based on a greedy algorithm [53] that performs a recursive binary partitioning of the feature space in order to build a decision tree. This algorithm uses the information gain in order to build the decision trees, and we used the default parameter as in the package *rpart* of R [54].

For the GBM, we used the GBM package of R [55] with Gaussian distribution, 3000 gradient boosting interactions, learning rate of 0.9 and 40 as maximum depth of variable interactions.

For RF, we used the implementation from provided by the randomForest package of R [56], using 100 as the number of trees to be built by algorithm and 100 as the maximum number of terminal nodes trees in the forest can have.

For ANN we used the nnet package of R [57], with maximum 10 number of hidden units, 10,000 maximum number of weights allowed and 1000 maximum number of iterations.

EV is an evolutionary algorithm for producing regression trees, and we used the R evtree package (from now on EVTree) [58], with parameters as in [14].

The ensemble method [14] uses a two layer strategy, where in the first layer random forests, neural networks and an evolutionary algorithm are used. The results produced by these three algorithms are then used by an algorithm based on Gradient Boosting in order to produce the final prediction.

All the parameters of the ML based techniques were established after several preliminary runs.

Table 5 shows the results obtained by the various methods for each value of w. We can notice that all the methods obtained better results with a historical window of 168 reads. NDL obtained the lowest MRE in all the cases, while the ensemble strategy obtains the second best results. Moreover, we can see that NDL outperforms all other methods even when only a historical window of 96 is used, confirming the extremely good performances of such strategy.

Table 5. Average results obtained by different methods for different historical window values. Standard deviation between brackets.

	w						
	24	**48**	**72**	**96**	**120**	**144**	**168**
NDL	3.01 (0.90)	2.38 (0.69)	2.08 (0.57)	1.85 (0.55)	1.60 (0.46)	1.51 (0.46)	1.44 (0.42)
CNN	4.08 (0.04)	3.16 (0.03)	2.69 (0.02)	2.51 (0.02)	2.30 (0.02)	1.71 (0.02)	1.79 (0.02)
LSTM	2.43 (0.03)	2.05 (0.02)	1.82 (0.02)	2.08 (0.02)	1.74 (0.02)	1.78 (0.02)	1.97 (0.02)
FFNN	4.51 (0.52)	3.46 (0.33)	3.39 (0.30)	3.12 (0.42)	2.98 (0.28)	2.32 (0.29)	2.46 (0.29)
ARIMA	8.82 (5.31)	8.26 (4.73)	11.37 (10.43)	14.03 (13.00)	6.79 (2.53)	7.63 (2.54)	6.92 (2.97)
DT	9.52 (1.55)	9.45 (1.48)	9.33 (1.39)	9.40 (1.45)	9.08 (1.12)	8.86 (1.01)	8.79 (0.96)
GBM	8.07 (3.82)	6.59 (2.71)	5.73 (2.23)	5.33 (2.08)	5.02 (1.81)	4.49 (1.54)	4.45 (1.56)
RF	4.39 (2.13)	3.69 (1.71)	2.93 (1.16)	2.78 (1.04)	2.45 (0.79)	2.22 (0.71)	2.15 (0.69)
EV	4.49 (1.91)	3.98 (1.52)	3.48 (1.18)	3.42 (1.15)	3.19 (0.95)	3.15 (0.90)	3.09 (0.84)
NN	4.39 (2.23)	4.27 (2.16)	4.13 (2.05)	3.55 (1.56)	3.15 (1.41)	2.16 (0.78)	2.08 (0.74)
ENSEMBLE	3.58 (1.65)	2.95 (1.19)	2.64 (0.99)	2.57 (0.97)	2.38 (0.81)	1.94 (0.69)	1.88 (0.67)

It is interesting also to notice that NDL obtains better results than DL for all the values of the historical window used, which confirms that using an evolutionary approach for optimizing the parameters of the deep learning network can be considered as a superior strategy with respect to grid optimization.

5. Conclusions and Future Works

In this paper, we proposed a strategy based on neuroevolution in order to predict the short-term electric energy demand. In particular, we used a genetic algorithm in order to obtain the architecture of a deep feed-forward neural network provided by the H2O big data analysis framework. The resulting networks have been applied to a dataset registering the electric energy consumption in Spain over almost 10 years.

The results were compared with other standard and machine learning strategies for time-series forecasting. For the experimentation performed we can conclude that the methodology we proposed in this paper is efficient for short-term electric energy forecasting, and on the particular dataset used in this paper the proposed strategy obtained the best performances. It is interesting to notice that our proposal outperforms the other ten strategies in all the cases, and that even when a historical window of 96 reads was used, our proposal achieved more precise predictions than any other methods with any other historical window size.

As for future work, we intend to apply the framework proposed in this paper to other datasets, and also to other kinds of time-series, in order to check the validity of our proposal also in other fields. Moreover, we intend to overcome a present limitation of the current proposal. In fact, the R GA package we have used does not allow to optimize parameter of different types, e.g., real and integer parameters. In order to overcome this, in this proposal we had to treat all the parameters as real. However, this causes the search space dimension to increase drastically. In the future we intend to solve this problem as well, and by reducing the size of the search space, we are confident that better configurations of the deep learning can be found. The use of on-line learning will also be explored in future works in order to speed up the prediction process and reduce the volume of stored data.

Author Contributions: F.D. conceived and partially wrote the paper. J.F.T. launched the experimentation. M.G.-T. and F.M.-Á. addressed the reviewers comments. A.T. validated the experiments. All authors have read and agree to the published version of the manuscript.

Funding: This research received no external funding.

Acknowledgments: The authors would like to thank the Spanish Ministry of Science, Innovation and Universities for the support under project TIN2017-88209-C2-1-R. This work has also been partially supported by CONACYT-Paraguay through Research Grant PINV18-661.

Conflicts of Interest: The authors declare no conflict of interest.

References

1. U.S. Energy Information Administration. International Energy Outlook. Available online: https://www.eia.gov/outlooks/ieo/index.php (accessed on 05 August 2020).
2. Narayanaswamy, B.; Jayram, T.S.; Yoong, V.N. Hedging strategies for renewable resource integration and uncertainty management in the smart grid. In Proceedings of the 3rd IEEE PES Innovative Smart Grid Technologies Europe, ISGT, Berlin, Germany, 14–17 October 2012; pp. 1–8.
3. Haque, R.; Jamal, T.; Maruf, M.N.I.; Ferdous, S.; Priya, S.F.H. Smart management of PHEV and renewable energy sources for grid peak demand energy supply. In Proceedings of the 2015 International Conference on Electrical Engineering and Information Communication Technology (ICEEICT), Dhaka, Bangladesh, 21–23 May 2015; pp. 1–6.
4. Kim, Y.; Son, H.; Kim, S. Short term electricity load forecasting for institutional buildings. *Energy Rep.* **2019**, *5*, 1270–1280. [CrossRef]
5. Nazeriye, M.; Haeri, A.; Martínez-Álvarez, F. Analysis of the Impact of Residential Property and Equipment on Building Energy Efficiency and Consumption-A Data Mining Approach. *Appl. Sci.* **2020**, *10*, 3589. [CrossRef]
6. Zekic-Suzac, M.; Mitrovic, S.; Has, A. Machine learning based system for managing energy efficiency of public sector as an approach towards smart cities. *Int. J. Inf. Manag.* **2020**, *54*, 102074. [CrossRef]
7. Energy 2020—A Strategy for Competitive, Sustainable and Secure Energy. Available online: http://eur-lex.europa.eu/legal-content/EN/TXT/PDF/?uri=CELEX:52010DC0639&from=EN (accessed on 5 August 2020).
8. Raza, M.Q.; Khosravi, A. A review on artificial intelligence based load demand forecasting techniques for smart grid and buildings. *Renew. Sustain. Energy Rev.* **2015**, *50*, 1352–1372. [CrossRef]
9. Torres, J.F.; de Castro, A.G.; Troncoso, A.; Martínez-Álvarez, F. A scalable approach based on deep learning for big data time series forecasting. *Integr. Comput.-Aided Eng.* **2018**, *25*, 1–14. [CrossRef]
10. Miikkulainen, R.; Liang, J.Z.; Meyerson, E.; Rawal, A.; Fink, D.; Francon, O.; Raju, B.; Shahrzad, H.; Navruzyan, A.; Duffy, N.; et al. Evolving Deep Neural Networks. *CoRR* **2017**, abs/1703.00548. Available online: https://arxiv.org/abs/1703.00548 (accessed on 5 August 2020) .
11. Stanley, K.O.; Clune, J.; Lehman, J.; Miikkulainen, R. Designing neural networks through neuroevolution. *Nat. Mach. Intell.* **2019**, *1*, 24–35. [CrossRef]
12. LeCun, Y.; Bengio, Y.; Hinton, G.E. Deep learning. *Nature* **2015**, *521*, 436–444. [CrossRef]
13. Such, F.P.; Madhavan, V.; Conti, E.; Lehman, J.; Stanley, K.O.; Clune, J. Deep Neuroevolution: Genetic Algorithms Are a Competitive Alternative for Training Deep Neural Networks for Reinforcement Learning. *CoRR* **2017**, abs/1712.06567. Available online: https://arxiv.org/abs/1712.06567 (accessed on 5 August 2020).
14. Divina, F.; Gilson, A.; Goméz-Vela, F.; Torres, M.G.; Torres, J.F. Stacking Ensemble Learning for Short-Term Electricity Consumption Forecasting. *Energies* **2018**, *11*, 949. [CrossRef]
15. Nowicka-Zagrajek, J.; Weron, R. Modeling electricity loads in California: ARMA models with hyperbolic noise. *Signal Process.* **2002**, *82*, 1903–1915. [CrossRef]
16. Huang, S.J.; Shih, K.R. Short-term load forecasting via ARMA model identification including non-Gaussian process considerations. *IEEE Trans. Power Syst.* **2003**, *18*, 673–679. [CrossRef]

17. Martínez-Álvarez, F.; Troncoso, A.; Asencio-Cortés, G.; Riquelme, J.C. A survey on data mining techniques applied to energy time series forecasting. *Energies* **2015**, *8*, 1–32. [CrossRef]
18. Muralitharan, K.; Sakthivel, R.; Vishnuvarthan, R. Neural network based optimization approach for energy demand prediction in smart grid. *Neurocomputing* **2018**, *273*, 199–208. [CrossRef]
19. Mordjaoui, M.; Haddad, S.; Medoued, A.; Laouafi, A. Electric load forecasting by using dynamic neural network. *Int. J. Hydrogen Energy* **2017**, *42*, 17655–17663. [CrossRef]
20. Wei, S.; Mohan, L. Application of improved artificial neural networks in short-term power load forecasting. *J. Renew. Sustain. Energy* **2015**, *7*, id043106. [CrossRef]
21. Gajowniczek, K.; Ząbkowski, T. Short Term Electricity Forecasting Using Individual Smart Meter Data. *Procedia Comput. Sci.* **2014**, *35*, 589–597. [CrossRef]
22. Min, Z.; Qingle, P. Very Short-Term Load Forecasting Based on Neural Network and Rough Set. In Proceedings of the Intelligent Computation Technology and Automation, International Conference on(ICICTA), Changsha, China, 11–12 May 2010; Volume 3, pp. 1132–1135.
23. Troncoso, A.; Riquelme, J.C.; Riquelme, J.M.; Martínez, J.L.; Gómez, A. Electricity Market Price Forecasting Based on Weighted Nearest Neighbours Techniques. *IEEE Trans. Power Syst.* **2007**, *22*, 1294–1301.
24. Martínez-Álvarez, F.; Troncoso, A.; Riquelme, J.C.; Aguilar-Ruiz, J.S. Energy time series forecasting based on pattern sequence similarity. *IEEE Trans. Knowl. Data Eng.* **2011**, *23*, 1230–1243. [CrossRef]
25. Shen, W.; Babushkin, V.; Aung, Z.; Woon, W.L. An ensemble model for day-ahead electricity demand time series forecasting. In Proceedings of the International Conference on Future Energy Systems, Berkeley, CA, USA, 22–24 May 2013; pp. 51–62.
26. Koprinska, I.; Rana, M.; Troncoso, A.; Martínez-Álvarez, F. Combining pattern sequence similarity with neural networks for forecasting electricity demand time series. In Proceedings of the IEEE International Joint Conference on Neural Networks, Dallas, TX, USA, 4–9 August 2013; pp. 940–947.
27. Jin, C.H.; Pok, G.; Park, H.W.; Ryu, K.H. Improved pattern sequence-based forecasting method for electricity load. *IEEJ Trans. Electr. Electron. Eng.* **2014**, *9*, 670–674. [CrossRef]
28. Wang, Z.; Koprinska, I.; Rana, M. Pattern sequence-based energy demand forecast using photovoltaic energy records. In Proceedings of the International Conference on Artificial Neural Networks, Nagasaki, Japan, 11–14 November 2017; pp. 486–494.
29. Bokde, N.; Asencio-Cortés, G.; Martínez-Álvarez, F.; Kulat, K. PSF: Introduction to R Package for Pattern Sequence Based Forecasting Algorithm. *R J.* **2017**, *1*, 324–333. [CrossRef]
30. Pérez-Chacón, R.; Asencio-Cortés, G.; Martínez-Álvarez, F.; Troncoso, A. Big data time series forecasting based on pattern sequence similarity and its application to the electricity demand. *Inf. Sci.* **2020**, *540*, 160–174. [CrossRef]
31. Zeng, B.; Li, C. Forecasting the natural gas demand in China using a self-adapting intelligent grey model. *Energy* **2016**, *112*, 810–825. [CrossRef]
32. Fan, G.F.; Wang, A.; Hong, W.C. Combining Grey Model and Self-Adapting Intelligent Grey Model with Genetic Algorithm and Annual Share Changes in Natural Gas Demand Forecasting. *Energies* **2018**, *11*, 1625. [CrossRef]
33. Ma, X.; Liu, Z. Application of a novel time-delayed polynomial grey model to predict the natural gas consumption in China. *J. Comput. Appl. Math.* **2017**, *324*, 17–24. [CrossRef]
34. Wu, Y.H.; Shen, H. Grey-related least squares support vector machine optimization model and its application in predicting natural gas consumption demand. *J. Comput. Appl. Math.* **2018**, *338*, 212–220. [CrossRef]
35. Martínez-Álvarez, F.; Asencio-Cortés, G.; Torres, J.F.; Gutiérrez-Avilés, D.; Melgar-García, L.; Pérez-Chacón, R.; Rubio-Escudero, C.; Troncoso, A.; Riquelme, J.C. Coronavirus Optimization Algorithm: A Bioinspired Metaheuristic Based on the COVID-19 Propagation Model. *Big Data* **2020**, *8*, 232–246. [CrossRef]
36. Torres, J.F.; Fernández, A.M.; Troncoso, A.; Martínez-Álvarez, F. Deep Learning-Based Approach for Time Series Forecasting with Application to Electricity Load. In *Biomedical Applications Based on Natural and Artificial Computing*; Springer International Publishing: Berlin, Germany, 2017; pp. 203–212.
37. Berriel, R.F.; Lopes, A.T.; Rodrigues, A.; Varejão, F.M.; Oliveira-Santos, T. Monthly energy consumption forecast: A deep learning approach. In Proceedings of the 2017 International Joint Conference on Neural Networks, IJCNN 2017, Anchorage, AK, USA, 14–19 May 2017; pp. 4283–4290.
38. Shi, H.; Xu, M.; Li, R. Deep Learning for Household Load Forecasting: A Novel Pooling Deep RNN. *IEEE Trans. Smart Grid* **2018**, *9*, 5271–5280. [CrossRef]

39. Guo, Z.; Zhou, K.; Zhang, X.; Yang, S. A deep learning model for short-term power load and probability density forecasting. *Energy* **2018**, *160*, 1186–1200. [CrossRef]
40. Talavera-Llames, R.L.; Pérez-Chacón, R.; Lora, A.T.; Martínez-Álvarez, F. Big data time series forecasting based on nearest neighbours distributed computing with Spark. *Knowl.-Based Syst.* **2018**, *161*, 12–25. [CrossRef]
41. Floreano, D.; Dürr, P.; Mattiussi, C. Neuroevolution: From architectures to learning. *Evol. Intell.* **2008**, *1*, 47–62. [CrossRef]
42. Kandasamy, K.; Neiswanger, W.; Schneider, J.; Póczos, B.; Xing, E. Neural Architecture Search with Bayesian Optimisation and Optimal Transport. *CoRR* **2018**, abs/1802.07191. Available online: https://arxiv.org/abs/1802.07191 (accessed on 5 August 2020).
43. Snoek, J.; Larochelle, H.; Adams, R.P. Practical Bayesian Optimization of Machine Learning Algorithms. In *NIPS'12, Proceedings of the 25th International Conference on Neural Information Processing Systems—Volume 2*; Curran Associates Inc.: New York, USA, 2012; pp. 2951–2959.
44. Assunção, F.; Lourenço, N.; Ribeiro, B.; Machado, P. Incremental Evolution and Development of Deep Artificial Neural Networks. In *Genetic Programming*; Hu, T., Lourenço, N., Medvet, E., Divina, F., Eds.; Springer International Publishing: Cham, Switzerland, 2020; pp. 35–51.
45. Assunção, F.; Lourenço, N.; Machado, P.; Ribeiro, B. Fast DENSER: Efficient Deep NeuroEvolution. In *Genetic Programming*; Sekanina, L., Hu, T., Lourenço, N., Richter, H., García-Sánchez, P., Eds.; Springer International Publishing: Cham, Switzerland, 2019; pp. 197–212.
46. Real, E.; Aggarwal, A.; Huang, Y.; Le, Q.V. Regularized Evolution for Image Classifier Architecture Search. *CoRR* **2018**, abs/1802.01548. Available online: https://arxiv.org/abs/1802.01548 (accessed on 5 August 2020). [CrossRef]
47. Real, E.; Moore, S.; Selle, A.; Saxena, S.; Suematsu, Y.L.; Tan, J.; Le, Q.V.; Kurakin, A. Large-Scale Evolution of Image Classifiers. In Proceedings of the 34th International Conference on Machine Learning, Sydney, Australia, 6–11 August 2017; Precup, D., Teh, Y.W., Eds.; PMLR: International Convention Centre: Sydney, Australia, 2017; Volume 70, pp. 2902–2911.
48. Spanish Electricity Price Market Operator. Available online: http://www.omie.es/files/flash/ResultadosMercado.html (accessed on 5 August 2020).
49. Team, T.H. H2O: R Interface for H2O. In *R Package Version 3.1.0.99999*; H2O.ai, Inc.: New York, NY, USA, 2015.
50. Scrucca, L. On some extensions to GA package: Hybrid optimisation, parallelisation and islands evolution. *R J.* **2017**, *9*, 187–206. [CrossRef]
51. Herrera, F.; Lozano, M.; Sánchez, A.M. A taxonomy for the crossover operator for real-coded genetic algorithms: An experimental study. *Int. J. Intell. Syst.* **2003**, *18*, 309–338. [CrossRef]
52. Salles, R.; Assis, L.; Guedes, G.; Bezerra, E.; Porto, F.; Ogasawara, E. A Framework for Benchmarking Machine Learning Methods Using Linear Models for Univariate Time Series Prediction. In Proceedings of the 2017 International Joint Conference on Neural Networks (IJCNN), Anchorage, AK, USA, 14–19 May 2017.
53. Rokach, L.; Maimon, O. Top-down Induction of Decision Trees Classifiers-a Survey. *Trans. Sys. Man Cyber Part C* **2005**, *35*, 476–487. [CrossRef]
54. Therneau, T.M.; Atkinson, B.; Ripley, B. rpart: Recursive Partitioning. Available online: https://rdrr.io/cran/rpart/ (accessed on 5 August 2020).
55. Ridgeway, G. Generalized Boosted Models: A Guide to the Gbm Package. Available online: https://rdrr.io/cran/gbm/man/gbm.html (accessed on 5 August 2020).
56. Liaw, A.; Wiener, M. Classification and Regression by randomForest. *R News* **2002**, *2*, 18–22.
57. Venables, W.N.; Ripley, B.D. *Modern Applied Statistics with S*, 4th ed.; Springer: New York, NY, USA, 2002.
58. Grubinger, T.; Zeileis, A.; Pfeiffer, K. evtree: Evolutionary Learning of Globally Optimal Classification and Regression Trees in R. *J. Stat. Softw.* **2014**, *61*, 1–29. [CrossRef]

Article

Analysis of the Impact of Residential Property and Equipment on Building Energy Efficiency and Consumption—A Data Mining Approach

Mahsa Nazeriye [1], Abdorrahman Haeri [1,*] and Francisco Martínez-Álvarez [2,*]

1 School of Industrial Engineering, Iran University of Science and Technology (IUST), Tehran 1684613114, Iran; ma_nazeriye@yahoo.com

2 Data Science & Big Data Lab, Pablo de Olavide University, ES-41013 Seville, Spain

* Correspondence: ahaeri@iust.ac.ir (A.H.); fmaralv@upo.es (F.M.-Á.)

Received: 14 February 2020; Accepted: 15 May 2020; Published: 22 May 2020

Abstract: Human living could become very difficult due to a lack of energy. The household sector plays a significant role in energy consumption. Trying to optimize and achieve efficient energy consumption can lead to large-scale energy savings. The aim of this paper is to identify the equipment and property affecting energy efficiency and consumption in residential homes. For this purpose, a hybrid data-mining approach based on K-means algorithms and decision trees is presented. To analyze the approach, data is modeled once using the approach and then without it. A data set of residential homes of England and Wales is arranged in low, medium and high consumption clusters. The C5.0 algorithm is run on each cluster to extract factors affecting energy efficiency. The comparison of the modeling results, and also their accuracy, prove that the approach employed has the ability to extract the findings with greater accuracy and detail than in other cases. The installation of boilers, using cavity walls, and installing insulation could improve energy efficiency. Old homes and the usage of economy 7 electricity have an unfavorable effect on energy efficiency, but the approach shows that each cluster behaved differently in these factors related to energy efficiency and has unique results.

Keywords: residential building; energy efficiency; clustering; decision tree

---

1. Introduction

In today's world, supplying energy is done through various carriers such as oil and gas (and products derived from them), electricity and renewable energy. Given the limited resources of energy and the population growth, the increasing annual consumption of energy affects life, the economy, the environment, politics, and so on. So, managing energy is a complicated task and has become an important issue in the modern world. The home section has the largest share of energy consumption in most countries. As each house has its own behavior, energy consumption patterns rely on several factors. Hence, decision making concerning the domestic sector's energy management and efficiency requires taking advantage of modern science capabilities to manage energy efficiency and consumption.

Data mining science can extract useful knowledge which is hidden in the data. Using its methods and algorithms, data mining techniques analyze huge amounts of data automatically [1,2]. In general, data mining is the process of Knowledge Discovery in Databases (KDD). Knowledge obtained from this fashionable science can be verified by conventional analysis [3]. As this science proves its potential in solving versatile problems [4], using it to extract knowledge in domestic buildings for managing energy is reasonable.

Scholars aim to develop innovation changes in the field of energy, so many studies have been conducted on the use of data mining science in energy management and efficiency. Energy performance certificates (EPCs) measure energy performance and give recommendations on how to improve energy efficiency [5,6]. They try to find inefficient building properties and improve them. In other words, they help to locate properties whose energy performance is effective and better them to improve their energy efficiency [7]. Pasichnyi et al. review existing applications of EPCs and present a method of EPC data quality assurance using data analytics [8]. Developments on EPCs have been done using data mining in recent years [9,10].

Some important insights related to the energy management requirement for buildings to save energy have been presented in [11,12]. Also, data mining tasks that can be used to mine building-related data have been shown in [13]. There are studies which focus on discovering factors which influence energy. Yan analyzed the impact of psychological, family and contextual aspects on residential energy consumption which indicate saving money, energy concern, and behavioral barriers which have a major impact on residential energy consumption behavior [14]. Effective factors on home electrical energy demand have been analyzed through developing a model using series prediction methods. The result demonstrates that houses using pool pumps and ducted air-conditioning have an increased electricity consumption, whereas houses with gas hot water systems have a lower power consumption than homes that do not use these systems [15]. The adaptive neuro-fuzzy inference system (ANFIS) has been used to discover major factors influencing energy consumption. This indicates that insulating materials are the most important parameters in building energy consumption. Attributes such as the type of materials and their thickness, wall structures, roofs and their ability to stay hot or cold, the location of walls and windows and geographic area have a major impact on energy saving [16]. The use of unsupervised learning has been applied to discover electricity consumption patterns in a Spanish public university. The authors found different clusters in which several buildings were identified. Such clusters were interpreted and rules for saving energy were proposed [17].

Clustering data is the process of putting data in a group so that they have the greatest similarity and are very dissimilar from data from other clusters. Many studies have been done in clustering [18–21]. Clustering is also used in the field of energy. The dataset has been classified into low, medium and high energy demand categories to show the factors influencing heating and hot water. A detailed analysis was performed using the k-means algorithm in the high consumption category. The output model presents good energy demand patterns and optimal ways to design buildings. The average U-value of the opaque envelope followed by the aspect ratio is the most important variable [22]. Characteristics of energy consumption examined have used cluster analysis for 134 LEED-NC certified office buildings. The buildings gathered into three clusters (low, medium and high) are very different and each one has a special attribute. The lower U-value of the roof and a lower ratio of windows are the factors which most influence a lower consumption. The HVAC system has a similar performance in all the clusters. The internal process load has a significant impact on clusters [23]. A framework based on data mining used CART classification and K-means clustering to analyze the pattern of energy consumption in a large data set of flats. Four influencing attributes (aspect ratio, U-value of vertical opaque envelope and windows, the average global efficiency of the system for space heating and DHW) were analyzed. High consumption flats were clustered and a reference flat was identified. These can be used to propose different energy retrofit actions [24].

The consumption of electricity and heating in six schools was studied using k-means clustering and self-organizing maps to evaluate energy efficiency. The schools have different construction years, areas, numbers of students and heating systems. Schools 1 and 4 have the highest cost of energy on working days and schools 2 and 3 have the lowest cost of energy at weekends compared to the others. The newest schools are generally better than the old ones in the field of energy efficiency [25]. A study of energy efficiency in 132 countries was estimated using a Data envelopment analysis model and then K-Means clustering, which specifies whether countries in a cluster are in the field of development or not. The results show that countries could develop energy efficiency by changing energy-related indicators [26]. A cluster analysis was used to analyze the regulations on energy efficiency of buildings of South America and Europe. This showed that buildings located in a similar climate zone but in different regions (countries) have different energy performances. It indicated that the tendencies of energy performance are different between various countries' regulations and the climate zones. The results confirmed the ability of cluster analysis to highlight similarity patterns between various regions of the same climate [27]. In the field of energy management systems, ISO 50001:2011, a systematic approach to improve the energy performance, plays an important role in the energy field. A study which classified, gathered, clustered and then applied data analysis techniques showed strategic decisions for improving the energy performance. The idea is used in an oil refinery and outputs of better energy management are shown [28]. Also in [29,30], efforts were made to develop the energy efficiency in industrial buildings. A fuzzy clustering technique was developed to rate school buildings in Greece. The methodology demonstrates that the energy consumption and global environmental quality of school buildings can be significantly improved, but the indoor air quality of these buildings causes some problems for them [31]. Wind is an energy source whose identification and assessment in its training needs is very important. The Analytic Hierarchy Process has been used to specify the training of wind farm employees. The results of the research prioritized the tasks and appropriate training courses tailored to the indicators were provided [32].

Discovering the rules is very much challenged in data mining [33–36]. Yu et al. present associations between building operational data. The methodology used on HVAC system data offers some "if-then" rules that are useful in the energy conservation field. Finding faulty equipment and repairing it, offering cost-efficient conservation strategies and a better understanding of building operation are suitable solutions for energy saving [37]. In another research work, the geographical and temperature variables in the electricity energy consumption were analyzed. Energy consumption and monthly average temperature data were clustered using K-means and then the Apriori algorithm was employed to discover association rules. These made "if-then" rules to describe the influence of different regions and physiographic objects. It shows that the most important parameters to increase electricity consumption are highways and then the ground, whereas rivers and farms (natural elements) decrease electricity consumption [38]. A combined framework using clustering and association rules developed to discover unusual energy used patterns. Benchmarking the rules identifies different waste patterns for different lifestyles [39]. A multi-objective algorithm is proposed to mine rules without a need to determine a minimum support threshold and a confidence threshold. This algorithm was used in three different datasets and it demonstrates its ability to mine quantitative association rules [40]. A hybrid algorithm including the genetic algorithm and particle swarm optimization algorithm was used to discover rules in continuous numeric datasets. It shows its ability in five different numerical interval datasets compared to other algorithms [41].

Due to irregular growth in the energy consumption of homes, analyzing their energy efficiency homes is an unavoidable study. Each building has its unique attribute, function and energy-related behavior to improve the energy efficiency of buildings. It is necessary to identify which factors and properties influence it, considering the unique behavior of homes. Analyzing them together leads to a tendency to pay attention to certain information while ignore others, and the findings are applicable to fewer buildings. This article proposes a hybrid approach that includes clustering and decision trees to identify factors affecting energy efficiency and consumption in residential buildings, as well as the reduction of the loss of some important data. The idea is that by clustering houses and putting similar patterns in a cluster, and then analyzing the factors in each cluster separately, findings will be extracted with more detail and accuracy than by not using the approach and analyzing them all together.

The rest of this paper is as follows: The next section provides the methodology used and the approach presented. In order to set forth the paper's purpose and also to examine the ability of the new approach, data analysis is done once without using the hybrid approach, provided in Section 3, and then again using the proposed approach in Section 4. Data clustering and also modeling each cluster separately will be done in this section. The evaluation and deployment are described in Section 5 and the conclusion is presented in Section 6, along with a discussion of the findings.

2. Methodology and Approach Presented

2.1. Methodology

Among the various methods of data mining science, such as [42], the Cross-Industry Standard Process for Data Mining (CRISP-DM) has been the one most widely used in data mining science. CRISP-DM is a global standard in project applications in data mining. This methodology consists of 6 phases, starting with the business understanding (problem definition) phase. The data understanding and the data preparation phases are done next. To achieve a basic understanding of the data, a cleaning and preparation of the data for modeling usage is done in these two phases. The modeling phase includes various techniques to analyze data and extract knowledge. The evaluation phase and then the deployment phase are the other phases of the CRISP-DM methodology [43]. In this methodology, phases could backtrack to previous phases. The SPSS Modeler of IMB [44] has been implemented with various tools and algorithms based on the CRISP-DM, The Clementine 12.0 released in Jan 2008 and IBM SPSS Modeler 18.0 released in March 2016 [45], software of IBM has been used to perform the data mining process.

Figure 1 shows the article methodology based on CRISP-DM. The article subject is defined in the first phase and it mainly discusses the problem definition. As expressed, the purpose is to identify properties affecting efficiency and also energy consumption in residential homes and also find out how to manage attributes and characteristics to achieve better energy management.

An understanding and preparation of the data is done in phase 2. A sample of 49,815 records of the housing stock of England and Wales has been selected. The Department of Energy and Climate Change has published this dataset [46]. Each record represented a region, a property age, a property type, the electricity and gas annual consumption from 2005 to 2012, the floor area band, etc. Table 1 describes the data set variables.

Table 1. Of variables.

Variable	Value	Description
HH_ID	1 to 49815	Household identifier.
REGION	(E12000001) North-East (E12000002) North-West (E12000003) Yorkshire and the Humber (E12000004) Mid-East (E12000005) Mid-West (E12000006) East England (E12000007) London (E12000008) South-East (E12000009) South-West (W999999999) Wales	Former Government Office Regions (GORs) in England, and Wales.
IMD_ENG	1 to 5	Index of multiple deprivations 2010 for England. Households are allocated to five groups. (1) The least deprived and (5) the most deprived.
IMD_WALES	1 to 5	Welsh Index of multiple deprivations 2011. This has five groups. (1) The most deprived and (5) the least deprived.
GconsYEAR		Annual gas consumption based on kWh.
GconsYEARValid		Flag of households' gas consumption. Valid gas consumption (V), households off the gas network (O) and invalid consumption.
EconsYEAR		Annual electricity consumption based on kWh.
EconsYEARValid		Flag indicating households with a valid electricity consumption.
E7Flag2012	(1) Households with Economy 7 electricity meters in 2012. (0) Households without Economy 7 electricity meters in 2012.	Economy 7 electricity meters.
MAIN_HEAT_FUEL	(1) gas (2) other	Main heating fuel.

Table 1. *Cont.*

Variable	Value	Description
PROP_AGE	(1) before 1930 (2) 1930 to 1949 (3) 1950 to 1966 (4) 1967 to 1982 (5) 1983 to 1995 (6) after 1996	Age of property construction.
PROP_TYPE	(1) detached (2) semi-detached (3) end-terrace (4) mid-terrace (5) bungalow (6) flat	Type of property.
FLOOR_AREA_BAND	(1) less than 50 m^2 (2) 51 m^2 to 100 m^2 (3) 101 m^2 to 150 m^2 (4) more than 150 m^2	Floor area band.
EE_BAND	(1) A and B (2) C (3) D (4) E (5) F (6) G	Energy Efficiency Band. (Six groups: A and B grouped).
LOFT_DEPTH	(1) less than 150 mm (2) 150 mm or more	Loft insulation depth.
WALL_CONS	(1) other (2) cavity wall	Wall construction.
CWI	(0) no (1) yes	Cavity wall insulation installed or not.
CWI_YEAR		Year of installation of cavity wall insulation.
LI	(0) no (1) yes	Loft insulation installed or not.
LI_YEAR		Year of installation of loft insulation
BOILER	(0) no (1) yes	Boiler installed in property or not.
BOILER_YEAR		Year of installation of the boiler.

The data is processed in phase 3. Discrepancy detection is done in this phase and there is also a negative impact on data quality which should be identified and resolved. The variable is O in 226 record values of the GconsYEARValid, which means that the household has not a gas network, while the values of the MAIN_HEAT_FUEL variable is 1, which means that the main heating fuel is gas. The records have been deleted due to a contradiction of the information. When the values of the GconsYEARValid variable is v, gas consumption must be between 100 kWh to 5000 kWh, but in 1107 records (2% of the records) the value of gas consumption when the GconsYEARValid variable is v is not valid so these 2% of records were deleted. Most values are the same in some variables. These variables did not affect the analysis and can be removed from the data set. GconsYEARValid and EconsYEARValid are variables with such a case. Data preparation/Modeling without a presented approach/Modeling using a proposed approach:

As the houses have different areas, different members, etc., the energy consumption can be different. To assess the electricity and gas consumption, these variables need to have a specific unit. So, the energy consumption has been normalized, based on the floor area (kWh/m^2). Since the exact area of each property is not available and the FLOOR_AREA_BAND variable is banded into four categories, the value of FLOOR_AREA_BAND variable is divided in the middle of each category of area variable to achieve a normal consumption based on kWh/m^2.

The data set has at times some variables which have no value in some records and there are no missing values. In other words, some features should have no values. So, these values are replaced to resolve the problem of blank values and because the algorithms do not consider these values the same as they do missing data. In this case, a value other than the value which is defined for that feature is set for these blank values for them not to be confused with missing values. The replacement is 0.514% of records. At the end of this phase, 48,898 refined records and 33 variables were obtained for the analysis.

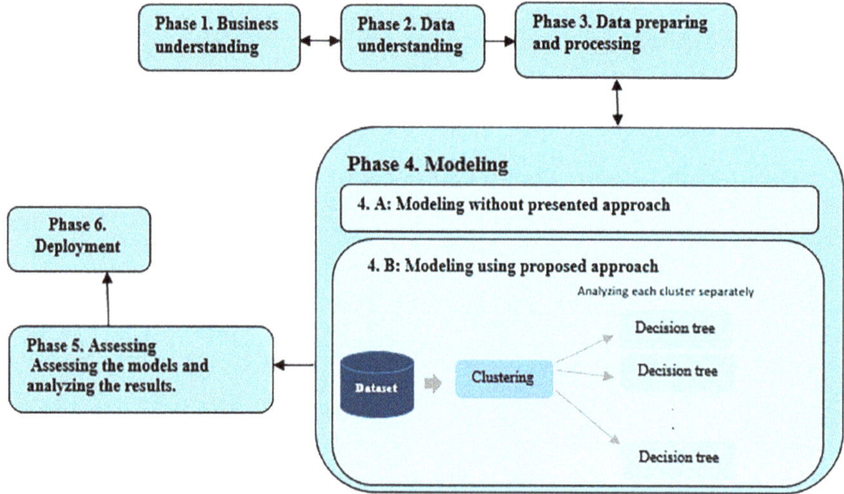

Figure 1. The methodology used—an overview.

The next phase, the modeling, simulated the prepared data obtained from phase 3 to extract knowledge and reveal the influence of property. This phase consists of two parts (Figure 1). First, modeling and analyzing the entire data altogether. Second, clustering data and then analyzing each cluster separately (the proposed approach). In fact, the goal is to identify how the results and findings using a combining approach and without using it differ and how the differences are effective in planning and decision making for the future. The first part (phase 4.A) is described in Section 3 and the second part (phase 4.B) will be described in detail in Section 4.

The models are then assessed to choose the most efficient model. In the evaluation phase, the knowledge gained from the previous phase evaluated whether the result of data analyzing could lead to the article's objectives or not. Also, the proposed approach's findings will be assessed to ensure that the approach presented in phase 4.B is able to provide more accurate knowledge. These two phases are described in detail in Section 3 to Section 4.

2.2. The Proposed Approach

Data mining is very powerful to discover unknowns in the absence of a prior knowledge of the data. Some minority records and their details are ignored, given that each record in the database has its unique attribute, and behavior modeling them all together causes data mining modeling and results which tend to yield a majority of records. By identifying records with similar patterns and grouping them in a cluster, and then analyzing each group separately, the results of the data mining tendency of a specific number of records, will be reduced to the minimum.

As each home has its unique attribute and property, analyzing all the data together yields a majority and some details are ignored. As shown in Figure 2, the idea is to put households with similar patterns (similar characteristics, attributes, and so on) in a cluster, and then evaluate the behavior and analyze the influential factors in each cluster separately. In this way, findings with more detail and accuracy are discovered. In fact, the article proposes a combined approach using the data mining technique with which data clustering will first be done and then each separate cluster will be modeled to identify more in depth the characteristics and factors influencing the energy efficiency.

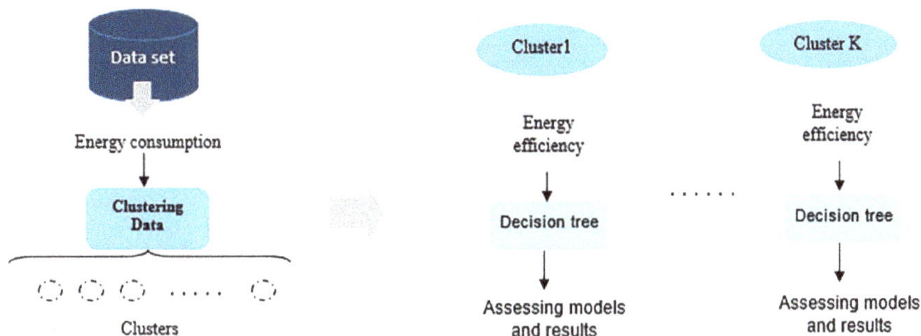

Figure 2. An overview of the proposed approach.

3. Modeling and Analyzing the Energy Efficiency without the Proposed Approach

As mentioned, the purpose of the paper is to discover the factors affecting energy efficiency and consumption and also to assess the proposed approach. So, the properties are modeled once without using the approach and then using it to analyze the energy efficiency in the domestic sector. The energy efficiency rate for each record obtained from EPCs logged for dwelling. The EPCs gather information on physical characteristics of the property and the main heating fuel, and gives score based on standard assumptions about residents and behavior. Then quantifies a dwelling's performance in terms of an efficiency rating (A the most efficient and G the less efficient). In this data set the most records' energy efficiency band is D (42.92%) and fewer records are in band A, B (2.71%) and G (1.31%).

This section deals with modeling and assessing the impact of properties on energy efficiency without the proposed approach. The aim of using decision trees is to obtain the most effective factor target class for each case in the data. For this purpose, the C5.0 algorithm [47] is used and all the variables except energy consumption are employed as the input, the energy efficiency group being the target. It can be said that the biggest advantage of C5.0 is that it presents its classification model as a tree structure which can be easily interpreted as rules. An advanced classifier may have better accuracy in many datasets but they cannot be easily understood and visualized. Also, the C5.0 reduces the pruning errors and has the ability of feature selection [48]. In C5.0, the root node is the most important variable and the best predictor. The leaf nodes contain a class label of the classification target.

The percentage presented in the tables show Ptrgt/Prule, which are:

- Ptrgt: Percentage of records that have the characteristics of the relevant rule and are also in the target energy efficiency group.
- Prule: Percentage of records that have the characteristics of the relevant rule.

Table 2 includes the results of analysis using the C5.0 algorithm. In this table, the rules corresponding to the C5.0 tree branches, which have a significant difference in the percentage of the target class, are presented in no particular order. It can be said that houses, which have a large share of better energy efficiency groups ((A, B) or C) are good cases for improving energy efficiency in other homes with similar attributes.

Table 2. The corresponding rules of the C5.0 tree (considering energy efficiency as the target).

Row	Rule	Result (Energy Efficiency)	Percentage
1	The built year is before 1930 and the home structure is semi-detached, has cavity walls, an insulation depth less than 150 mm and uses economy 7 electricity meters.	G	75%
2	The home area is 51 m^2 to 100 m^2 and has a mid-terrace structure. The built year is 1966 and 1982, the home has cavity walls and its boiler was installed in 2009.	C	100%
3	The home area is 51 m^2 to 100 m^2 and has a mid-terrace structure. The built-year is 1966 and 1982, the home has cavity walls and its boiler was installed in 2010.	F	100%
4	The home area is 51 m^2 to 100 m^2 and has a mid-terrace structure. The built-year is 1966 and 1982, the home has cavity walls and its boiler was installed in 2006 or 2011.	D	80%
5	The home area is 51 m^2 to 100 m^2 and has a mid-terrace structure. The built year is 1966 and 1982, the home has cavity walls and its boiler was installed in 2005 or 2012.	E	71.50%
6	The region is 4 and the built year is 1930 to 1949, the home structure is a flat and which has cavity walls.	C	100%
7	The region is 4 and the built year is 1930 to 1949, the home structure is a flat and it does not have cavity walls.	D	66.70%
8	The region is 1 and the home structure is semi-detached, and the built year is after 1966. The homes have not installed wall insulation, and they use economy 7 electricity meters.	D	100%
9	The region is 1 and the home structure is semi-detached, and the built year is after 1966. The homes have not installed wall insulation, and they do not use economy 7 electricity meters.	C	66.80%
10	The houses are in Wales and their built-year is before 1930, the home structure is semi-detached and does not have cavity walls, their insulation depth is less than 150 mm, and a boiler has been installed.	C	100%
11	The region is 1 and the built year is before 1930, the home structure is semi-detached, and does not have cavity walls, and a boiler has been installed and their insulation depth is less than 150 mm	F	100%
12	The region is 3 and the built-year is before 1930, the home structure is semi-detached, and does not have cavity walls, the home's boiler was installed in 2009 and loft insulation was used.	D	100%
13	The region is 3 and the built-year is before 1930, the home structure is semi-detached, and does not have cavity walls, the home's boiler was installed in 2009 and loft insulation was not used.	E	66.80%

According to the rules, some knowledge can be discovered.

- Carefully scrutinizing the rules of rows 2 to 5, it is concluded that installing a boiler will result in better energy efficiency. Dwelling which install boilers in 2009 are in better energy efficiency group. They are followed by the boilers installed in 2005 and 2012. The homes whose boiler was installed in 2010 have not a good Energy Efficiency.
- A comparison of rules 8 and 9 indicates that in area 1, tariffs do not yield an improved energy efficiency. Of course, this was seen among other households, but was not provided due to the low support value.
- In the rules of rows 1, 11 and 13, the household energy efficiency group is very bad. The issue is their built year (built before 1930). Obviously, in old houses, the thermal performance and energy consumption of equipment are weak compared to new ones. In general, only 0.35% of the old homes of datasets have energy efficiency groups A and B, while nearly 52% of them are in weak energy efficiency groups (E, F, and G). But in newly-built houses (after 1996), these percentages reach 13.8% for the energy efficiency groups A and B, and only 2% for poor energy efficiency groups.
- Rules 1, 10 and 11 refer to similar old houses that are located in different regions. Among these rules, the homes of Wales are in better condition in terms of energy efficiency.
- Data set records are in different climate zones (Table 1) and same energy efficiency rating is not obtained with similar conditions in a cold climate zone or a warm climate zone. Region 7 has the most A, B rating (3.84%) and region 5 the least (1.99%). This difference should be referring to the buildings structure, different equipment, and surely the family lifestyle.

- It is obvious that installing insulation on the ceiling and walls leads to a reduction of energy dissipation. Rules 12 and 13 show that an improvement in energy efficiency is achieved by installing insulation in the roof of residential buildings. Also, rules 6 and 7 state that the structure of the cavity wall is better than other structures. Policies to install new insulators in homes that do not have the proper equipment, especially among older homes, can lead to significant improvements in energy efficiency.

In general, old houses have a very bad energy efficiency. Installing proper insulation and using appropriate wall structure, also using equipment with energy efficiency grade of A or B can be effective in improving the efficiency of these homes. The installation of boilers and the non-use of electricity tariffs have led to better energy efficiency among households of this dataset. Various regions of England and Wales have more desirable homes in terms of energy efficiency than the rest.

4. Modeling and Analyzing the Energy Efficiency Using the Proposed Approach

4.1. Clustering Data

Each home has different characteristics and energy consumption. Categorizing data based on the author's opinion and the distribution of features is not very appropriate because it involves the author's assumptions and speculation. In this type of category, the probability of error and inaccuracy increases, which is contrary to the purpose of the article, to achieve results with greater accuracy. Cluster analysis is an unsupervised learning technique which finds data that has the most similarity with each other, and also the greatest difference with other data, and places them in a group called a cluster.

Clustering algorithms have a wide range that can be named partitioning, hierarchical, density-based, and grid-based methods. The residential buildings of this article are clustered using the k-means algorithm. This algorithm is used for clustering in different data sets [49] and also in energy consumptions field for different datasets [50–52]. Among different indicators for estimating the optimal number of clusters [53,54], the silhouette index [55] has been selected to calculate with a different number of clusters. The silhouette has a range of −1 to 1, where 1 indicates the best matched and −1 indicates variables which are poorly matched to their cluster.

Table 3 shows the Silhouette index values of clustering data of this article. While this indicator is an important factor for clustering, it should be noted that in the real world and information retrieval, clusters must have a comprehensible interpretation (cluster labeling). So, the selection of the best number of clusters should be based on a combination of the index and the labeling. The silhouette value of 2 and 3 clusters is greater than others (Table 3), which means that these clusters have a better coherence, although these values are close together. Therefore, the appropriate value is that which has a better interpretation and labeling adequate to the cluster's data attributes. Regarding the values of variables, in either case, three clusters have more interpretation and make a better differentiation within and between clusters. Hence, it was selected as the best number of clusters.

Table 3. Silhouette index values.

Number of Clusters	Silhouette Index Value
2	0.243
3	0.239
4	0.189
5	0.195
6	0.155
7	0.155
8	0.166
9	0.158
10	0.179

The three-clusters clustering results are as follows. Figure 3 indicates the size of these clusters. According to these characteristics and the average annual energy consumption of each cluster, cluster 1, cluster 2 and cluster 3 are labeled as medium-consumption, high-consumption and low-consumption clusters.

- Cluster 1: This cluster's homes are old (almost 68% built before 1930). Most of the households have a D label in the energy efficiency group and half of them have houses with an area of 51 m^2 to 100 m^2.
- Cluster 2: 27.5% of the homes were built between 1968 and 1982. Most of the households have a D label in the energy efficiency group and 58.6% of them have houses with an area of 51 m^2 to 100 m^2. The households of this cluster have more energy consumption than those of the other clusters.
- Cluster 3: Most of the households have a C label in the energy efficiency group. The cluster homes are small (59% of homes have an area less than 51 m^2). In comparison to other clusters, this cluster contains more newly-built houses and these also have the lowest energy consumption.

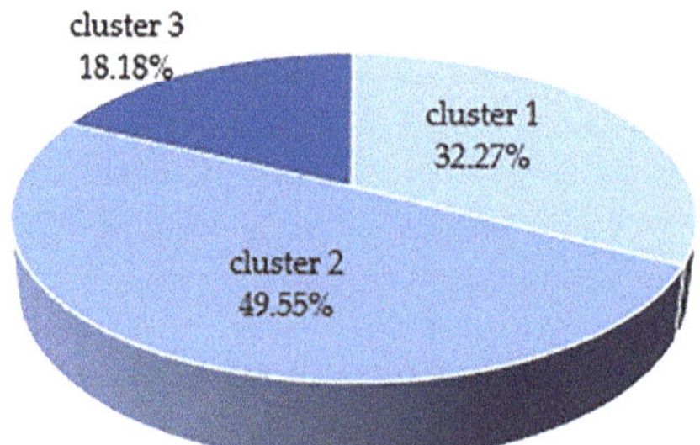

Figure 3. Size of clusters.

4.2. Modeling the Energy Efficiency in Each Cluster

According to the approach presented, each cluster which includes records with the most similarity must be analyzed separately to identify factors which influence the energy efficiency group. The corresponding rules of the decision trees' (C5.0) branches in the separate analysis of each cluster are given in Table 4. The percentage presented in the tables is explained in Section 3.

The findings of Table 4 show that:

In the Low-consumption cluster:

- Economic tariff 7 Electricity in this cluster has also shown its impact. More than 34% of the homes that have this tariff have the D label, while more than 46% of the homes which do not use this tariff are labeled C. By comparing rules 2 and 3, it can also be concluded that the use of this tariff in small houses has a greater impact on poor energy efficiency.
- Survey households in this cluster also state that a greater percentage of homes that do not use this electricity tariff 7 are better in energy efficiency than those using tariff 7. This different percentage in the small houses of this cluster (area less than 51 m^2) is shown in Table 5.
- Rules 4 and 5 stated that newly built houses are in good condition in terms of energy efficiency and old ones have a poor energy efficiency. The survey reveals that 36.9% of households living in newly built houses have an A or B label, and 3.28% of the old ones have a G label.

Table 4. The corresponding rules of the C5.0 tree in each cluster separately.

Cluster Label	Row	Rule	Result (Energy Efficiency)	Percentage
Low-consumption cluster	1	The home structure is a bungalow and uses economy 7 electricity meters and has installed a boiler.	D	80%
	2	The home structure is mid-terrace and the built-year is between 1950 and 1966, the floor area is less than 51 m^2 and it uses economy 7 electricity meters.	F	75%
	3	The home is in region 2 and the floor area is between 51 m^2 and 100 m^2, the built-year is between 1950 and 1966 and it uses economy 7 electricity meters.	D	71.50%
	4	The home structure is detached houses built before 1930.	G	66.70%
	5	Homes built after 1996, having cavity walls, and whose floor area is between 51 m^2 to 100 m^2 and which do not use economy 7 electricity meters.	A and B	54.90%
Medium-consumption cluster	6	The home structure is detached and was built before 1930, does not have a cavity wall, and the floor area is more than 151 m^2, it uses economy 7 electricity meters.	E	100%
	7	The region is Wales, and the home structure is semi-detached and built before 1930 and has a boiler installed.	C	100%
	8	The home structure is detached and built before 1930 and does not have a cavity wall, the floor area is more than 151 m^2 and it does not use economy 7 electricity meters.	F	80%
	9	The home structure is a flat built after 1996, does not have a cavity wall and does not use economy 7 electricity meters.	A and B	67.90%
	10	The home structure is detached, semi-detached, mid-terrace, and end-terrace, built after 1996, does not have a cavity wall and does not use economy 7 electricity meters.	C	65.70%
High-consumption cluster	11	The home structure is detached and the region is Wales and it was built between 1983 and 1995, a boiler was installed in 2010 and the home does not use economy 7 electricity meters.	A and B	100%
	12	The home structure is detached and the region is Wales and it was built between 1983 and 1995, a boiler was installed in 2010 and the home uses economy 7 electricity meters.	D	100%
	13	The home structure is a flat, built after 1996 and the floor area is between 51 m^2 and 100 m^2.	A and B	100%
	14	Houses built after 1996, and the structure is mid-terrace.	C	81.40%
	15	The home structure is detached and the region is 4, 5 and 9, it was built in 1983 to1995, a boiler was installed in 2010 and it does not use economy 7 electricity meters.	C	80%
	16	The home structure is mid-terrace, it was built between 1983 and 1995 and has a boiler installed.	C	78.80%
	17	Houses built after 1996; the structure is semi-detached.	C	72.50%
	18	The home structure is mid-terrace, it was built between 1966 and 1982 and has boilers installed.	C	72.20%

Table 5. The percentages of homes located in different energy efficiency groups.

Household Characteristic	Energy Efficiency Group					
	A, B	C	D	E	F	G
The small homes which have tariff 7	6.92%	40.34%	32.54%	13.51%	5.47%	1.50%
The small homes which do not have tariff 7	12.74%	51.23%	26.51%	6.73%	2.23%	0.68%

In the medium-consumption cluster:

- Rules 6 and 8 stated that, firstly, large old houses have not a good energy efficiency. Also, they state that among these homes, the ones which use electricity tariff 7 are better off.
- Rules 9 and 10 refer to the role of house structures in energy efficiency. So, in newly-constructed houses with a flat structure the energy efficiency groups are A and B, and other house structures have a C label.

- In Wales, houses which have a detached structure, are old (built before 1930) and which have installed boilers have an energy efficiency group C (rule 7).

In the high-consumption cluster:

- The good performance of boiler installation in the residential houses of this cluster is evident. In the 12th, 16th and 17th rules, the homes are not newly-built, but are located in the energy efficiency group C. The existence of a boiler in these houses, structured as mid-terrace and end-terrace, has diminished the role of the built-year.
- A comparison of rules 11, 15 and, 18 shows that similar homes in different areas have different energy efficiencies. The study of climatic conditions shows that the groups of areas that are located in one rule are not similar in terms of temperature and have different climates. Thus, it cannot be said that just the similarity or the difference in weather has led to different energy efficiencies. However, in this cluster, households living in Wales have generally a better energy efficiency.
- In the rules 13 and 14, newly-built houses are noted in good energy efficiency groups. It is important to say that newly-built flats have the best efficiency. In fact, the type of house structure is not ineffective in the energy performance.

5. Assessment and Deployment

5.1. Assessment

Phase 5 of the methodology measures the models evaluated in the previous sections and the accuracy of modeling. Using training and testing sets is an important step to evaluate the decision tree accuracy. The higher the accuracy of the model means that its performance is better. So, the data were divided into two groups (training and testing set). 70% of the records were taken as the training set and the remaining 30% as the testing set. In addition to accuracy, the lift criterion, which indicates the degree of correlation, was calculated for the C5.0 tree branches. If its value is greater than 1, this means a positive correlation. For all the branches presented in Tables 2 and 4, the lift value was more than one, hence indicating a positive correlation.

As stated, the purpose of the proposed approach is to achieve results with more precision and details. Therefore, the accuracy of the models of Sections 3 and 4 should be compared in order to assess the efficiency and effectiveness of this approach. In Table 6 the accuracy of the C5.0 tree in all the data and in each cluster is provided. As can be seen, the accuracy of modeling using the proposed approach (modeling in each cluster separately) is greater than analyzing all the data together. This means that the presented approach exposed unknown information more accurately and its tendency to a majority of records has declined.

Since this classification is of a multi-class type, the target feature (energy efficiency group) has six different modes ((A and B), C, D, E, F and G), the percentages obtained being acceptable. If the accuracy were by chance, as the target field has six different states, the accuracy of the tree would be about 17% (1/6). But Table 4 offers percentages other than this. The percentages indicate that the accuracy of the algorithm in the analysis of each cluster separately is greater than the total analysis of the data. This approach is capable of providing knowledge and discovering unknown information more accurately.

Table 6. Accuracy of the C5.0 Algorithm in the entire data and also in each cluster.

Input Data	The C5.0 Accuracy
All the data	54.4%
Cluster 1 (medium-consumption)	51.5%
Cluster 2 (high-consumption)	57.83%
Cluster 3 (low-consumption)	69.3%

Data clustering is also evaluated through the silhouette index (Table 3). Based on the values of this index in different scenarios and the analysis of characteristics in different values of k, 3 clusters were selected as the most suitable number of clusters (mentioned in Section 4.1)

5.2. Deployment and Discussion

Leading research has been conducted to explore the factors affecting energy efficiency in residential homes. To this end, a hybrid approach was proposed to reveal findings with greater detail and accuracy. This section reviews the findings. These suggest that some were commonly found in modeling without using the proposed approach and using it. These findings are as follows.

- In general, the installation of boilers will lead to an improved energy efficiency. Dwelling which installed boilers in 2009 have better energy efficiency than others and ones which install boiler in 2010 have the weakest performance, which is seen as an urgent need to replace or modify these boilers.
- Most old homes suffer from unfavorable energy efficiency. This is also reflected in the proposed approach. Homes in the high-consumption cluster are older than those in the low-consumption cluster. In old houses, the appliances and structures have a poor energy efficiency performance. Therefore, planning for structural improvements, installing proper insulation and switching equipment, especially in high-consumption cluster houses, is a constructive way to improve energy efficiency.
- Not using electricity tariffs 7 yields better energy efficiency group in most homes.

The interesting point is that the proposed approach, in addition to the results described above, could also reveal new findings. In fact, this approach offers more detailed results (Table 7). A scrutiny of the data through the proposed approach provides new findings, as follows.

- In homes with similar attributes, not using this tariff has resulted in a better energy efficiency group. However, the electricity tariff 7 has different effects in each cluster. An analysis of the approach presented shows that in the low-consumption cluster, the energy efficiency is poor, particularly in small (less than 51 m^2) and old houses which do not use this tariff.

It was especially seen in the medium-consumption cluster that big (over 151 m^2) and old houses which use this tariff have a better energy efficiency.

- The building structure influences different effects in the proposed approach. In the medium-consumption cluster, flats have a more favorable energy efficiency group than other structures, even among the newly built ones. Also, old homes which are structured as detached have a good energy efficiency group in this cluster. On the other hand, it has been seen before that old houses do not have a good energy efficiency.

In the high-consumption cluster, buildings with mid-terrace and end-terrace structures which have installed boilers belong to a better energy efficiency group.

- In the high-consumption cluster, homes in Wales have better energy efficiency than homes with similar attributes but which are in different areas.

Table 7 shows the findings, analyzing the entire dataset and each cluster separately. New findings extracted which are obtained from the proposed approach are shown as new finding. This suggests that the approach presented can discover findings in more detail and this proves the ability and usefulness of this approach in discovering unknown information. These detailed findings can be very helpful for policy maker, architects, and engineers.

Table 7. Assessing the results of the proposed approach.

Evaluated Data	Findings and Results			
All Data (without using the approach)	Installing boilers lead to better performance.	Old homes (built before 1930) suffer from unfavorable energy efficiency. Old homes in Wales have better energy efficiency.	Not using tariff 7 leads to better energy efficiency.	Improving energy efficiency is achieved by installing insulation in the roof of residential buildings.
Low-consumption cluster			Not using tariff 7 leads to better energy efficiency. Using the tariff leads to poor energy efficiency in small and old houses (less than 51 m^2). (new finding)	
Medium-consumption cluster		Flats have a more favorable energy efficiency; even newly-built flats. (new finding) Old homes in Wales which have a detached structure, and have installed boilers have a good energy efficiency. (new finding)	Using this tariff leads to a better energy efficiency in big (over 151 m^2) and old houses. (new finding)	
High-consumption cluster	Installing boilers leads to a better energy efficiency. In mid-terrace and end-terrace structures, boilers lead to better energy efficiency. (new finding)	The newly constructed flats have the best efficiency. (new finding)		Households living in Wales have better energy efficiency. (new finding)

6. Conclusions

The excessive demand for energy is a major challenge for countries. Therefore, governments seek to improve energy management and efficiency to reduce energy waste. In this article, a hybrid approach based on clustering and classification proposes discovering factors affecting energy efficiency in the domestic sector.

49,815 examples of the housing stock of England and Wales were used. First, households were analyzed to identify the influence of factors using a decision tree (without using the proposed approach). Then, the proposed approach was used. The K-means algorithm yields three clusters (low-consumption, medium-consumption, and high-consumption clusters). Households in each cluster were analyzed using the C5.0 algorithm. Comparing the results and modeling accuracy, once without using the approach and then using it, showed the ability of the approach presented to identify the properties that affect energy efficiency and consumption in-depth and more accurately. The approach presented is adaptable to different data sets.

The results of not using the approach shown is that installing boilers has improved the energy efficiency, especially with respect to those that dwelling installed in 2009, followed by boilers installed in 2005 and 2012. Dwelling install boilers 2010 have the least performance. Using electricity tariff 7 results in poor efficiency. Also, in old homes the thermal equipment and energy consumption of the equipment are weak compared to newly-built houses, which results in in weaker energy efficiency. The data also showed that walls which have a cavity structure as well as insulating installation lead to improved energy efficiency. Of course, the cavity wall itself has different types, which produces comments on how these walls affect the energy efficiency. The need for information on the type of cavity wall used in the residential buildings is urgent to find out more thorough knowledge.

Besides the findings presented above, the proposed approach provides new and more detailed results. It is demonstrated that electricity tariff 7 has different behaviors in different clusters. Generally, it was seen that the use of this tariff is not good to improve energy efficiency. The approach shows that in the low-consumption cluster, old and small houses (less than 51 m^2) that use this tariff have a poor energy efficiency. Also, among the old and big houses (over 151 m^2) of the medium-consumption cluster using this tariff has a positive impact.

The approach shows that the home structure influences energy efficiency. In the high-consumption cluster, installing boilers in mid-terrace and end-terrace structures, and detached structures in Wales in the medium-consumption cluster, leads to better energy efficiency. In the medium-consumption cluster, it was seen that flats are better in energy efficiency than other structures, even compared to newly-built houses.

Different geographic regions also had a different behavior. The high-consumption cluster shows better energy efficiency in the houses in Wales. Definitely, having more comprehensive and adequate information of the different regions of England and Wales could extract more knowledge. The accuracy of modeling in the approach presented was better than modeling without it and detailed findings can be discovered.

Through its new and in-depth results, this approach has shown that it is capable and beneficial in the field of retrieval knowledge. These new findings demonstrate that we cannot make a similar decision for all homes. As homes in a cluster have their unique behavior, policies and decisions must be unique for them. The knowledge obtained is suitable and useful for residential buildings of similar features and nature to plan and upgrade energy efficiency, and also to improve EPCs.

Author Contributions: M.N. and A.H. proposed the idea; M.N. designed the model and the computational framework, also carried out the implementation and processed the experimental data, performed the analysis and designed the figures, interpreted the results and wrote the manuscript, performed the proofreading, discussed the results; A.H. verified the analytical methods and supervised the findings of this work, discussed the results; F.M.-Á. verified the analytical methods, discussed the proofreading, contributed in discussing the results. All authors have read and agreed to the published version of the manuscript.

Funding: This research received no external funding.

Acknowledgments: The authors thank the reviewers for their valuable suggestions for improving the manuscript. This research did not receive any specific grant from funding agencies in the public, commercial, or not-for-profit sectors.

Conflicts of Interest: The authors declare no conflict of interest.

References

1. Han, J.; Pei, J.; Kamber, M. *Data Mining Concepts and Techniques*, 2nd ed.; Elsevier: Amsterdam, The Netherlands, 2006.
2. Martínez-Álvarez, F., Troncoso, A., Asencio-Cortés, G.; Riquelme, J.C. A survey on data mining techniques applied to energy time series forecasting. *Energies* **2015**, *8*, 1–32. [CrossRef]
3. Read, B.J. Data mining and science? Knowledge discovery in science as opposed to business. In Proceedings of the 12th ERCIM Workshop on Database Research, Amsterdam, The Netherlands, 2–3 November 1999.
4. Yu, Z.; Fung, B.C.; Haghighat, F. Extracting knowledge from building-related data—A data mining framework. In *Building Simulation*; Tsinghua Press: Beijing, China, 2013; Volume 6, pp. 207–222.
5. Watts, C.; Jentsch, M.F.; James, P.A. 'Evaluation of domestic Energy Performance Certificates in use'. *Build. Serv. Eng. Res. Technol.* **2011**, *32*, 361–376. [CrossRef]
6. Watson, P. An introduction to UK Energy Performance Certificates (EPCs). *J. Build. Apprais.* **2010**, *5*, 241–250. [CrossRef]
7. Di Corso, E.; Cerquitelli, T.; Piscitelli, M.S.; Capozzoli, A. Exploring Energy Certificates of Buildings through Unsupervised Data Mining Techniques. In Proceedings of the 2017 IEEE International Conference on Internet of Things (iThings) and IEEE Green Computing and Communications (GreenCom) and IEEE Cyber, Physical and Social Computing (CPSCom) and IEEE Smart Data (SmartData), Exeter, UK, 21–23 June 2017; pp. 991–998.
8. Pasichnyi, O.; Wallin, J.; Levihn, F.; Shahrokni, H.; Kordas, O. Energy performance certificates — New opportunities for data-enabled urban energy policy instruments? *Energy Policy* **2019**, *127*, 486–499. [CrossRef]

9. Koo, C.; Hong, T. Development of a dynamic operational rating system in energy performance certificates for existing buildings: Geostatistical approach and data-mining technique. *Appl. Energy* **2015**, *154*, 254–270. [CrossRef]
10. Liu, J.; Wang, J.; Li, G.; Chen, H.; Shen, L.; Xing, L. Evaluation of the energy performance of variable refrigerant flow systems using dynamic energy benchmarks based on data mining techniques. *Appl. Energy* **2017**, *208*, 522–539. [CrossRef]
11. IEEE Industry Applications Society. *Power Systems Engineering Committee. IEEE Recommended Practice for Electric Power Systems in Commercial Buildings*; American National Standards Institute: New York, NY, USA, 1991.
12. Danish, M.S.S.; Senjyu, T.; Ibrahimi, A.M.; Ahmadi, M.; Howlader, A.M. A managed framework for energy-efficient building. *J. Build. Eng.* **2019**, *21*, 120–128. [CrossRef]
13. Yu, Z.J.; Haghighat, F.; Fung, B.C. Advances and challenges in building engineering and data mining applications for energy-efficient communities. *Sustain. Cities Soc.* **2016**, *25*, 33–38. [CrossRef]
14. Yan, S.L. Influence of psychological, family and contextual factors on residential energy use behavior: An empirical study of China. *Energy Procedia* **2011**, *5*, 910–915. [CrossRef]
15. Fan, H.; MacGill, I.F.; Sproul, A.B. Statistical analysis of driving factors of residential energy demand in the greater Sydney region, Australia. *Energy Build.* **2015**, *105*, 9–25. [CrossRef]
16. Naji, S.S. Application of adaptive neuro-fuzzy methodology for estimating building energy consumption. *Renew. Sustain. Energy Rev.* **2016**, *53*, 1520–1528. [CrossRef]
17. Pérez-Chacón, R.; Luna, J.M.; Troncoso, A.; Martínez-Álvarez, F.; Riquelme, J.C. Big data analytics for discovering electricity consumption patterns in smart cities. *Energies* **2018**, *11*, 683. [CrossRef]
18. Kao, Y.T.; Zahara, E.; Kao, I.W. A hybridized approach to data clustering. *Expert Syst. Appl.* **2008**, *34*, 1754–1762. [CrossRef]
19. Niknam, T.; Amiri, B.; Olamaei, J.; Arefi, A. An efficient hybrid evolutionary optimization algorithm based on PSO and SA for clustering. *J. Zhejiang Univ. A* **2009**, *10*, 512–519. [CrossRef]
20. Fathian, M.; Amiri, B. A honey-bee mating approach on clustering. *Int. J. Adv. Manuf. Technol.* **2007**, *38*, 809–821. [CrossRef]
21. Laszlo, M.; Mukherjee, S. A genetic algorithm that exchanges neighboring centers for k-means clustering. *Pattern Recognit. Lett.* **2007**, *28*, 2359–2366. [CrossRef]
22. Capozzoli, A.G. Discovering knowledge from a residential building stock through data mining analysis for engineering sustainability. *Energy Procedia* **2015**, *83*, 370–379. [CrossRef]
23. Heidarinejad, M.D. Cluster analysis of simulated energy use for LEED certified U.S. office buildings. *Energy Build.* **2014**, *85*, 86–97. [CrossRef]
24. Capozzoli, A.; Serale, G.; Piscitelli, M.S.; Grassi, D. *Data Mining for Energy Analysis of a Large Data Set of Flats*; Thomas Telford Ltd.: London, UK, 2017.
25. Raatikainena, M.S.-P. Intelligent analysis of energy consumption in school buildings. *Appl. Energy* **2016**, *165*, 416–429. [CrossRef]
26. Jalali Sepehr, M.; Haeri, A.; Ghousi, R. A cross-country evaluation of energy efficiency from the sustainable development perspective. *Int. J. Energy Sector Manag.* **2019**, *13*, 991–1019. [CrossRef]
27. Bienvenido-Huertas, D.; Oliveira, M.; Rubio-Bellido, C.; Marín, D. A Comparative Analysis of the International Regulation of Thermal Properties in Building Envelope. *Sustainability* **2019**, *11*, 5574. [CrossRef]
28. Haeri, A.; Rezaie, K. An approach to evaluate resource utilization in energy management systems. *Energy Sources Part B* **2016**, *11*, 855–860. [CrossRef]
29. Haeri, A. Proposing a quantitative approach to measure the success of energy management systems in accordance with ISO 50001: 2011 using an analytical hierarchy process (AHP). *Energy Equip. Syst.* **2017**, *5*, 349–355. [CrossRef]
30. Haeri, A.; Jafari, M.; Danesh Asgari, S. A new approach for performance evaluation of energy-related enterprises. *Energy Equip. Syst.* **2018**, *6*, 16–26. [CrossRef]
31. Santamouris, M.; Mihalakakou, G.; Patargias, P.; Gaitani, N.; Sfakianaki, K.; Papaglastra, M.; Pavlou, C.; Doukas, P.; Primikiri, E.; Geros, V. Using intelligent clustering techniques to classify the energy performance of school building. *Energy Build.* **2007**, *39*, 45–51. [CrossRef]
32. Haeri, A. Identification and assessment of training needs for employees of wind farms'. *Energy Equip. Syst.* **2017**, *5*, 189–196. [CrossRef]

33. Huang, M.-J.; Sung, H.-S.; Hsieh, T.-J.; Wu, M.-C.; Chung, S.-H. Applying data-mining techniques for discovering association rules. *Soft Comput.* **2019**, *24*, 8069–8075. [CrossRef]
34. Zhang, C.; Xue, X.; Zhao, Y.; Zhang, X.; Li, T. An improved association rule mining-based method for revealing operational problems of building heating, ventilation and air conditioning (HVAC) systems. *Appl. Energy* **2019**, *253*, 113492. [CrossRef]
35. Li, G.; Hu, Y.; Chen, H.; Li, H.; Hu, M.; Guo, Y.; Liu, J.; Sun, S.; Sun, M. Data partitioning and association mining for identifying VRF energy consumption patterns under various part loads and refrigerant charge conditions. *Appl. Energy* **2017**, *185*, 846–861. [CrossRef]
36. Fan, C.; Xiao, F. Mining Gradual Patterns in Big Building Operational Data for Building Energy Efficiency Enhancement. *Energy Procedia* **2017**, *143*, 119–124. [CrossRef]
37. Yu, Z.; Jerry, H.F. A novel methodology for knowledge discovery through mining associations between building operational data. *Energy Build.* **2012**, *47*, 430–440. [CrossRef]
38. Rathod, R.R.; Garg, R.D. Regional electricity consumption analysis for consumers using data mining techniques and consumer meter reading data. *Electr. Power Energy Syst.* **2016**, *78*, 368–374. [CrossRef]
39. Li, J.; Panchabikesan, K.; Yu, Z.; Haghighat, F.; Mankibi, M.; Corgier, D. Systematic data mining-based framework to discover potential energy waste patterns in residential buildings. *Energy Build.* **2019**, *199*, 562–578. [CrossRef]
40. Moslehi, F.; Haeri, A. A Genetic Algorithm based framework for mining quantitative association rules without specifying minimum support and minimum confidence. *Sci. Iran.* **2019**. [CrossRef]
41. Moslehi, F.; Haeri, A.; Martínez-Álvarez, F. A novel hybrid GA–PSO framework for mining quantitative association rules. *Soft Comput.* **2019**, *24*, 4645–4666. [CrossRef]
42. Shafique, U.; Qaiser, H. A comparative study of data mining process models (KDD, CRISP-DM and SEMMA). *Int. J. Innov. Sci. Res.* **2014**, *12*, 217–222.
43. Wirth, R.; Hipp, J. CRISP-DM: Towards a standard process model for data mining. In Proceedings of the 4th International Conference on the Practical Applications of Knowledge Discovery and Data Mining, Manchester, UK, 11–13 April 2000.
44. IBM SPSS Modeler-Data Mining, Text Mining, Predictive Analysis. Available online: http://www.spss.com.hk/software/modeler/ (accessed on 22 March 2020).
45. SPSS Predictive Analytics. Announcing IBM SPSS Modeler 18 - SPSS Predictive Analytics. 2016. Available online: https://developer.ibm.com/predictiveanalytics/2016/03/15/announcing-ibm-spss-modeler-18/ (accessed on 18 May 2020).
46. National Energy Efficiency Data-Framework (NEED): Anonymised Data 2014. Available online: https://www.gov.uk/government/statistics/national-energy-efficiency-data-framework-need-anonymised-data-2014 (accessed on 18 May 2020).
47. Quinlan, J.R. Induction of decision trees. *Mach. Learn.* **1986**, *1*, 81–106. [CrossRef]
48. Pandya, R.; Pandya, J. C5.0 algorithm to improved decision tree with feature selection and reduced error pruning. *Int. J. Comput. Appl.* **2015**, *117*, 18–21.
49. Moslehi, F.; Haeri, A.; Gholamian, M. A novel selective clustering framework for appropriate labeling of the clusters based on K-means algorithm. *Sci. Iran.* **2019**. [CrossRef]
50. Gaitani, C.L. Using principal component and cluster analysis in the heating evaluation of the school building sector. *Appl. Energy* **2010**, *87*, 2079–2086. [CrossRef]
51. Hsu, D. *Characterizing Energy Use in New York City Commercial and Multifamily Buildings*; ACEEE Summer Study on Energy Efficient in Buildings: Pacific Grove, CA, USA, 2012.
52. Xiao, Q.W.H. The reality and statistical distribution of energy consumption in office buildings in China. *Energy Build.* **2012**, *50*, 259–265. [CrossRef]
53. Dunn, J.C. *A Fuzzy Relative of the ISODATA Process and Its Use in Detecting Compact Well-Separated Clusters*; Taylor & Francis: Abingdon, UK, 1973.

54. Davies, D.L.; Bouldin, D.W. A cluster separation measure. *IEEE Trans. Pattern Anal. Mach. Intell.* **1979**, *2*, 224–227. [CrossRef]
55. Rousseeuw, P.J. Silhouettes: A graphical aid to the interpretation and validation of cluster analysis. *J. Comput. Appl. Math.* **1987**, *20*, 53–65. [CrossRef]

Article

Stability of Multiple Seasonal Holt-Winters Models Applied to Hourly Electricity Demand in Spain

Óscar Trull [1], J. Carlos García-Díaz [1] and Alicia Troncoso [2,*]

1 Department of Applied Statistics, Operational Research and Quality, Universitat Politècnica de València, 46022 Valencia, Spain; otrull@eio.upv.es (Ó.T.); juagardi@eio.upv.es (J.C.G.-D.)

2 Division of Computer Science, Universidad Pablo de Olavide, ES-41013 Seville, Spain

* Correspondence: atrolor@upo.es

Received: 21 February 2020; Accepted: 6 April 2020; Published: 10 April 2020

Abstract: Electricity management and production depend heavily on demand forecasts made. Any mismatch between the energy demanded with respect to that produced supposes enormous losses for the consumer. Transmission System Operators use time series-based tools to forecast accurately the future demand and set the production program. One of the most effective and highly used methods are Holt-Winters. Recently, the incorporation of the multiple seasonal Holt-Winters methods has improved the accuracy of the predictions. These forecasts, depend greatly on the parameters with which the model is constructed. The forecasters need to deal with these parameters values when operating the model. In this article, the parameters space of the multiple seasonal Holt-Winters models applied to electricity demand in Spain is analysed and discussed. The parameters stability analysis leads to forecasters better understanding the behaviour of the predictions and managing their exploitation efficiently. The analysis addresses different time windows, depending on the period of the year as well as different training set sizes. The results show the influence of the calendar effect on these parameters and if it is necessary or not to update them in order to obtain a good accuracy over time.

Keywords: time series; forecasting; exponential smoothing; electricity demand

---

1. Introduction

The programming of the production of electrical energy is a complex task carried out by the Transmission System Operators (TSO). Their main objective is the supply of electricity, assuring the distribution. In addition, the energy production at the lowest possible cost to the consumer, without incurring losses is a key point [1]. Any mismatch between the programmed and the really demanded energy produces huge losses. Hobbs [2] determined that the losses produced by a 1% gap between planning and reality can cost up to millions of dollars. Hong et al. [3] determined that for a maximum peak demand central of 1 GW, a 1% error in the prediction can involve costs of $600,000 per year. The TSOs use for their predictions time series forecasting [4–6]. One of the most common used technique is the exponential smoothing methods, and especially the Holt-Winters models, due to their easy to understand and implement features [7,8]. Exponential smoothing methods use the data observed in the past to make predictions, assigning an exponentially decreasing weight to the older information against the newer. The way to assign the relevance of newer data against the older is performed by the weight assigned to smoothing parameters. The smoothing parameters are bounded in the range [0,1]. The closer to 0 the more important the older data is, and the contrary when closer to 1. The Holt-Winters model is described in Equations (1)–(4)

$$S_t = \alpha \left(\frac{X_t}{I_{t-s}} \right) + (1-\alpha)(S_{t-1} + T_{t-1}) \tag{1}$$
$$T_t = \gamma(S_t - S_{t-1}) + (1-\gamma)T_{t-1} \tag{2}$$
$$I_t = \delta \left(\frac{X_t}{S_t} \right) + (1-\delta)I_{t-s} \tag{3}$$
$$\widehat{X}_{t+h} = (S_t + hT_t)I_{t-s+h} \tag{4}$$

where S_t, and T_t are the equations for the level and additive trend, with smoothing parameters α and γ. I_t are the seasonal indices of length s, with smoothing parameters δ. X_t is the observed data. Finally, $\widehat{X}_{t+h}$ is the equation to forecast h time instants ahead. It collects the information contained in the model and makes predictions.

The introduction of double and triple seasonal models (HWT), described in [9,10], improved the accuracy of these methods, and their use is being generalized. Additionally, these methods behave very accurate for such type of series, with a strong seasonal effect [11,12]. The generalization to include up to n seasonal patterns is known as Multiple Seasonal Holt-Winters (nHWT), developed in [13]. The model is defined as in Equations (5)–(8),

$$S_t = \alpha \left(\frac{X_t}{\prod_{i=1}^{n_s} I_{t-s_i}^{(i)}} \right) + (1-\alpha)(S_{t-1} + T_{t-1}) \tag{5}$$
$$T_t = \gamma(S_t - S_{t-1}) + (1-\gamma)T_{t-1} \tag{6}$$
$$I_t^{(i)} = \delta^{(i)} \left(\frac{X_t}{S_t \prod_{j=1, j\neq i}^{n_s} I_{t-s_j}^{(j)}} \right) + (1-\delta^{(i)})I_{t-s_i}^{(i)} \quad i = 1, \ldots, n_s \tag{7}$$
$$\widehat{X}_{t+h} = (S_t + hT_t) \prod_{i=1}^{n_s} I_{t-s_i+h}^{(i)} + \varphi_{AR}\varepsilon_t \tag{8}$$

where $I_t^{(i)}$ are the seasonal indices of length s_i, with smoothing parameters $\delta^{(i)}$. There are as many equations as seasonal patterns allocated in the time series, n_s. The φ_{AR} parameter is entered to correct the model including effectively the first order autocorrelation error (ε_t), known as AR(1) adjustment.

Depending on the way the equations are defined, additive or multiplicative methods can be used. The method of combining the smoothing equations will determine the Holt-Winters model to be used. The combination provides 30 different models, according to the trend and seasonality combination, as well as AR(1) adjustment, as described in [13]. The nomenclature for these models is as follows: Three letters to describe the combination, where the first one describes the trend method, the second one the seasonality method and the last one whether the adjustment was applied. This combination is shown in Table 1. Furthermore, the model is described by adding as subscript the different seasonal patterns considered, using the length of the cycle. As an example, the $AMC_{24,168}$ one is a model with additive trend and multiplicative seasonal, the model is adjusted using first order autocorrelation error and it has two seasonalities, one daily (subscript 24) and one weekly (subscript 168).

The smoothing parameters are obtained by adjusting the model to data observed in the past, and try to reproduce the same pattern for the near future. The adjustment is done by solving a non-linear problem, in which the 1-hour-ahead forecasting error is minimised. This topic is better developed in Section 3. The smoothing parameters obtained are assumed to be optimal, and the model is then exploited. The optimality of these parameters is locally obtained, valid only for the dataset and the model selected. Moreover, the same parameter can highly vary from one method to another using the same dataset. Thus, it is a big deal for forecasters to understand the behaviour of the model considering the locally optimised parameters. The future predictions accuracy will be closely related to the smoothing parameters. The optimisation procedure is using an optimisation algorithm, that do not take care concerning the reality of the series. The forecaster needs to make use of the own experience to validate and approve the obtained parameters. These actions are crucial for the entire process. Some factors have an influence on the parameter values. The method to obtain initial values for seeding the model impact on the forecasting accuracy [14] although after an optimisation of the parameters, accuracy differences are also minimised [15]. Climate conditions are not a big deal in terms of short-term forecasting accuracy [12]. However, it seems reasonable the parameters ought to be influenced. If only two seasonal patterns are considered, seasons modify the trend, whereas if three seasonalities are considered, the intra-year seasonal parameter should be influenced. The calendar also has a huge influence on the predictions and parameters, that must deal with some irregularities of the series [16–18]. Therefore, forecasters should always pay close attention to the parameters values.

Table 1. Multiple seasonal Holt-Winters models' nomenclature according to trend and seasonal method. The first letter determines the trend, N: none, A: additive, d: damped additive, M: multiplicative and D: damped multiplicative. The second letter is used for Seasonality, with only N (none), A (additive) and M (multiplicative) option. The third letter is used for the AR(1) adjustment: L (no adjustment) and C (adjusted).

Trend \ Seasonality	None	Additive	Multip.	None	Additive	Multip.
		Normal			AR(1) Adjusted	
None	NNL	NAL	NML	NNC	NAC	NMC
Additive	ANL	AAL	AML	ANC	AAC	AMC
Damped additive	dNL	dAL	dML	dNC	dAC	dMC
Multiplicative	MNL	MAL	DML	MNC	MAC	MMC
Damped multiplicative	DNL	DML	DML	DMC	DAC	DMC

The forecasting procedure performed by TSOs must ensure continuously accurate forecasts for the electricity system. These predictions are used by the TSO for the operational planning and unit assignment, while they are also used by the market for the spot price settlement [19,20]. The Spanish electricity market (OMIE) operates similarly [21,22], where the unique Spanish TSO, Red Eléctrica de España (REE), supplies forecasts of demand for the next week every Wednesday, for the next 24 h every day, and a revision of the daily demand every six hour [23]. The market uses this information for the bidding process, as well as REE itself to carry out operational planning. Thus, there is an enormous responsibility in such eagerly awaited forecasts. In fact, REE uses a complex algorithm [24] to provide forecasts.

With such high frequency forecasting, there is a need for the model to be updated continuously. The forecaster must deal with the varying smoothing parameters and take decisions on the engaged forecasts. In the Holt-Winters literature there is a strong discussion about whether the parameters should be continuously readjusted as the series progresses, or on the contrary, they must be immobile and try to exploit them for as long as possible. The values of the parameters are also discussed, since high values of the parameters indicate a great damping to adapt to the new observed values, compared to low values, where the initial values are given greater weight. Before this doubt, the following question is added: how do the parameters respond to a specific time series, such as the hourly electricity demand in Spain? Make it sense the obtained values? What is their stability? All these questions have not been answered in the previous literature.

This article describes how we have performed a stability analysis on the parameters of the nHWT models applied to the electricity demand series in Spain. The main objective was to understand the behaviour of smoothing parameters against several different datasets with changes in climate conditions, time series components such as trend or seasonal, calendar effects among others. This analysis will help the forecasters to face the optimised model before making forecasts, considering the values of the parameters. It is also important to check whether it is necessary or not to update the smoothing parameters of the Holt-Winters models in order to obtain a good accuracy over time. On the other hand, the results of this study will show the influence of the calendar effect on these parameters, and therefore the need to develop new Holt-Winters models that take it into account [16].

The article is organized as follows: Section 2 reviews the existing literature related to the smoothing parameters of Holt-Winters models. Section 3 presents the methodology followed in order to predict the Spanish electricity demand time series and to analyze the variability of the parameters; in Section 4 the results obtained for a given double and triple seasonal model are shown and in Section 5 the results are discussed. Finally, the conclusions reached in this article are shown.

2. Related Work

Short and medium term electricity demand time series forecasting were extensively studied in the literature applying both statistical methods and machine learning techniques. Recently, the machine learning methods have focused on big data [25], especially deep learning [26,27], and ensemble methodologies [28,29].

Within classical methods, the accuracy of the forecasts in the Holt-Winters models has been widely studied, especially how to select the best method and adjust the parameters [30]. As the Holt–Winters models are recursive, an initialization value is needed to feed the model. Thus, the initialization methods for Holt-Winters were also another intense field of research especially with the emergence of double and triple seasonal Holt-Winters models [31,32]. In fact, new methods to initialize the level, trend and seasonality in multiple seasonal Holt–Winters models were recently developed in [14]. These seed values have influence on the smoothing parameters' values as well as the forecasting accuracy [33]. However, when the series adjustment is made, the initial values lose influence [15]. After fitting the Holt-Winters model, the obtained smoothing parameters are assumed to be optimal for the data set. Thus, the study of their values or their stability has not been worked in depth in the literature. This is due, in large part, to the fact that the models do not allow a theoretical mathematical analysis.

Archibald [34] used 406 monthly series from the M competitions. He analyzed the models with additive seasonality, and demonstrated that the values of the parameters within the range [0,1] are not always invertible. Thus, it is necessary to use only a set range within the region of invertibility.

Lawton [35] used state spaces to analyze Holt-Winters models, and although his work focused on the normalization of the seasonal component, he also analyzed the stability of the parameters. From previous work on filters [36,37], it determined that Holt-Winters models are not asymptotically stable. The values of the eigenvectors of the stability matrices depend on the α, γ and δ parameters.

Some authors [38] worked on obtaining the limits of the smoothing parameters. They analyzed state space parameters, and stated the parameters ought to meet $0 < \alpha < \gamma$. Finally, the authors in [39] established a series of criteria in the parameters so that the models can be "predictable", a term that mints whether a series will be able to make forecasts with constant mean and variance. Osman et al. [40] obtained the values of the main vectors and confirmed that the models proposed in [39] produce stable predictions, but if regressors are used, they must be invariant over time.

Another interesting study related to this paper is the study of the minimum size required for the sample. Hyndman et al. [41] analyzed this situation and concluded that there is no clear answer, but that there should be many more observations than parameters. In this regard, García-Díaz and Trull [13] verified that for a double seasonal model, a data set length of one year produces more stable parameters than 8–10 weeks.

3. Materials and Methods

The time series used in this study is the hourly electricity demand in Spain, within the years from 2008 to 2017. It was provided by REE through its website www.ree.es. This time series shows clearly several seasonalities: the intraday which is repeated every 24 h, and the intra-weekly, which is repeated every 168 h. A third seasonality has also been included, which relates the data of the series with the seasonal periods of the year. Various strategies were proposed to deal with the 3rd seasonality, since its length depends largely on the concept to be covered. Taylor uses the calendar year, so that the years comprise 52 weeks, although to adjust the process in leap years, it uses 53 weeks [10]. Most authors have chosen to use the solar year, which includes 365.25 days, and seasonally adjusted [42]. In our case, it was considered to be 365.25 days, or 8766 h.

This series was used in former works [43,44] and especially using the nHWT methods [13,14,16]. This experience showed the authors that many factors could have an impact on the model parameters, as well as its forecast accuracy, key point for the proper exploitation of the nHWT models. The influence of the initial values was analysed in [14]. However, there is still a lack of analysis regarding the meaning of the parameter space and their relation with forecasts.

The method described in this paper analyses the smoothing parameters of the nHWT forecasting procedure when working with the hourly electricity demand in Spain. It performs an analysis of the parameter values, by checking their behaviour depending on the method chosen from Table 1. That implies to analyse the stability against some influential factors derived from our experience such as seasons, years, calendar, and inner variability.

To match the previously mentioned issues, raw data must be used, avoiding any smoothing or modification of the series. Double seasonal nHWT used a dataset size of 8–10 weeks, it was chosen as this seems to be long enough for this purpose. Alternatively, an analysis of the dataset size is also performed. Triple seasonal nHWT used a dataset of three years.

Randomness in the series selection must be assured. To observe the maximum possible variability, randomly selected data sets were used, but always complying with the following premises:

- At least 3 sets of data for each season of the year.
- At least there must be sets in different years, to check the repeatability of the process in similar situations in a matter of period of the year, including holidays nearby.
- The data has not been filtered or the special days were altered.

As a result, for the seasonal double analysis, 250 random samples of 10 weeks were extracted, of which, the first 8 weeks were used for adjustment, while the other two weeks were used for validation. Because the climatic effect present in the series must be assessed, the samples were selected so that they belong to different climatic periods and several years. Figure 1 shows the working scheme for this analysis.

The parameters optimization was done using a minimization algorithm, Nelder-Mead's simplex [45], where the minimization function was the Root Mean of the Squared Error (RMSE), described in Equation (9).

$$RMSE = \sqrt{\frac{1}{h}\sum_{t=1}^{h}(\widehat{X}_t - X_t)^2} \tag{9}$$

where h is the number of samples to be predicted.

Once the model was fit and the parameters obtained, 24-hours ahead forecast during two weeks were performed, and compared against the real data. The forecast accuracy was measured by using the Mean Average Percentage Error (MAPE), described in Equation (10).

$$MAPE(\%) = 100\frac{1}{h}\sum_{t=1}^{h}\frac{|\hat{X}_t - X_t|}{|X_t|} \tag{10}$$

The use of RMSE for the model optimisation is preferred [46]. On the contrary, the most commonly used indicator to compare forecast accuracy is MAPE [47], in special when comparing demand forecasts. The triple seasonal analysis was performed similarly, but it was needed a higher dataset to fit the model and obtain the parameters. Thus, 80 random samples from three full years and two weeks were sampled. The first set of three years was used to fit the model, whereas the two week set was used for validation purposes. In this case, only the $AMC_{24,168,8766}$ model was used for the analysis.

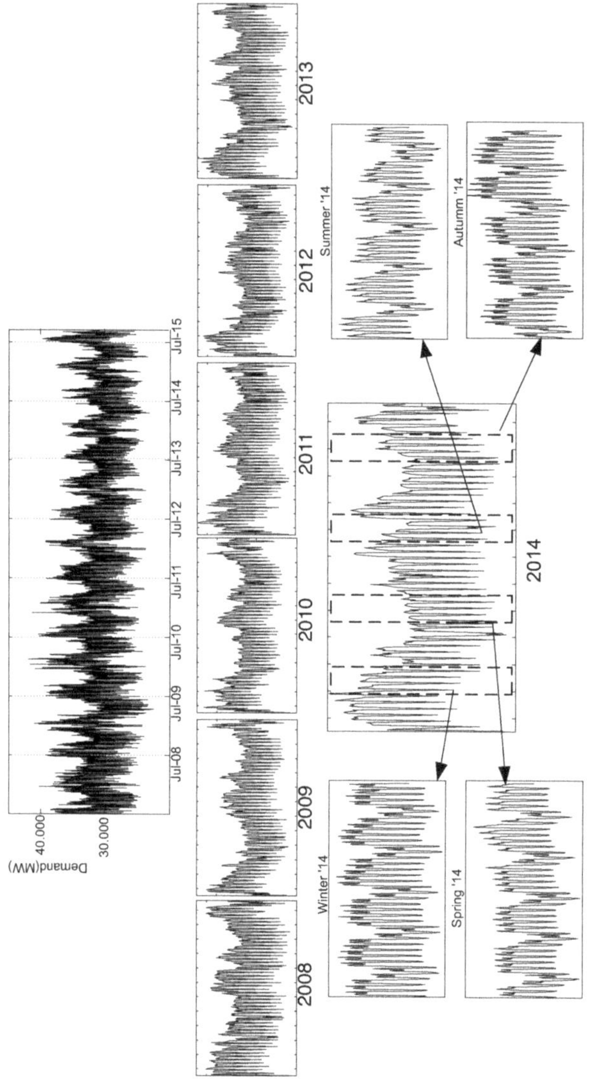

Figure 1. Demand split in years and seasons for stability analysis.

4. Results

The analysis procedure was organized inductively. First, the results of the forecasts and how they and the smoothing parameters evolve over time were observed for both double and triple seasonal Holt-Winters models, and then, the influence of the size of the adjusting set was analyzed.

4.1. Double Seasonal Models

Table 2 shows the results of the forecasts made by double seasonal models. In particular, the average of the MAPE when predicting the 250 random samples of two weeks. Only models with the correction of the first order autocorrelation error setting are shown, since they offer better results. Removing the year 2009, the year in which the crisis was emphasized in the Spanish industrial sector until that uncertain moment, the rest of the years there is a clear stability in the forecasts.

Holt-Winters models are very robust to small variations. The forecast accuracy values are around 2.6% in terms of MAPE for 24 hours-ahead forecasts. In the triple seasonal case these values are around 7.9% of MAPE.

Table 2. Summary of the 24-hour ahead forecasting MAPE of double seasonal models, split by years.

Year	2008	2009	2010	2011	2012	2013	2014	2015	2016	2017	Mean
$AAC_{24,168}$	2.4755	3.3322	2.8705	2.5367	2.6761	2.2657	2.5447	2.4527	2.6741	2.6238	2.6963
$AMC_{24,168}$	2.4644	3.2576	2.8389	2.4974	2.5697	2.2284	2.4225	2.4131	2.7129	2.6972	2.6521
$dAAC_{24,168}$	2.4402	3.3052	2.8227	2.5120	2.5927	2.2272	2.5114	2.4397	2.6361	2.6189	2.6571
$dAMC_{24,168}$	2.4028	3.1723	2.8047	2.4720	2.5209	2.2190	2.3866	2.4164	2.6447	2.5585	2.6034
$DMAC_{24,168}$	2.4483	3.3217	2.8034	2.5183	2.6054	2.2351	2.4846	2.4279	2.6399	2.6296	2.6601
$DMMC_{24,168}$	2.4119	3.1572	2.8087	2.4519	2.5267	2.2045	2.4077	2.4117	2.6330	2.6184	2.6026
$MAC_{24,168}$	2.4722	3.3320	2.8427	2.5260	2.6059	2.2430	2.5031	2.4420	2.6520	2.6325	2.6746
$MMC_{24,168}$	2.4462	3.1844	2.8538	2.4670	2.5681	2.2444	2.4787	2.5781	2.7370	2.7615	2.6531
$NAC_{24,168}$	2.4434	3.3208	2.8131	2.4970	2.5932	2.2258	2.4869	2.4256	2.6497	2.6216	2.6552
$NMC_{24,168}$	2.4061	3.1339	2.8043	2.4196	2.5126	2.1720	2.4407	2.3999	2.6366	2.5810	2.5872

The $NMC_{24,168}$ is selected as it offers the best forecast accuracy. As the MAPE provided is an average each year, a split including the months is graphed in Figure 2 as weather conditions and months may influence on the results. Clearly the winter and autumn months have higher levels of MAPE, this time the winter months when worse forecasts occur.

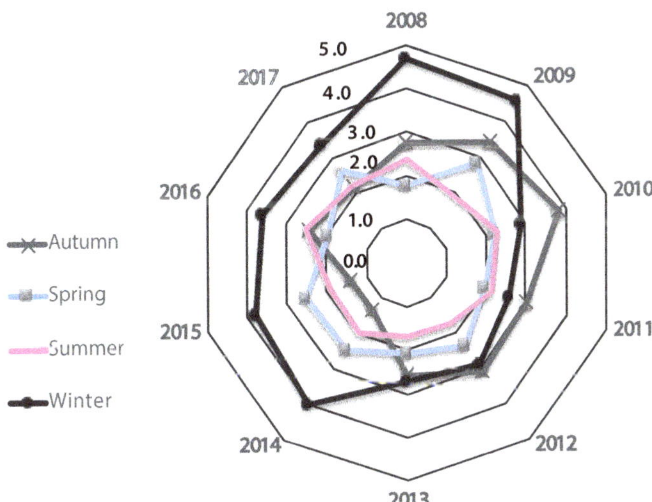

Figure 2. MAPE radar diagram of the forecasts for the $NMC_{24,168}$ model using the demand time series from 2008 to 2017.

Table 3 shows the distribution of the parameter values according to the year and seasonal period for the model $NMC_{24,168}$. These values were obtained using the sets for adjustment composed of the 250 random samples of 8 weeks. The total row is the mean of the parameters if they are calculated without dividing into seasons. It can be seen that there was a shift in the parameter values from autumn to winter. The values for the daily seasonality are in the order of 0.2 to 0.3 with the exception of winter, where it increases to the value of 0.4. For the intra-weekly seasonality it has a value of 0.2 and it becomes 0.5 for the winter season.

Table 3. Distribution of the parameters for the $NMC_{24,168}$ model.

		α		$\delta^{(24)}$		$\delta^{(168)}$		φ_{AR}	
		mean	**Dev.**	**mean**	**Dev.**	**mean**	**Dev.**	**mean**	**Dev.**
	2009								
Autumn		0.0647	0.0398	0.3003	0.0588	0.2660	0.0795	0.8716	0.0487
Spring		0.0099	0.0074	0.3762	0.1326	0.1668	0.0316	0.9480	0.0245
Summer		0.0502	0.0288	0.2407	0.0870	0.2044	0.0637	0.8938	0.0420
Winter		0.6009	0.4141	0.4240	0.2209	0.6386	0.3875	0.6615	0.1899
Total		0.1623	0.2982	0.3331	0.1457	0.3048	0.2548	0.8522	0.1390
	2010								
Autumn		0.0805	0.0610	0.2890	0.0281	0.2152	0.0422	0.9011	0.0446
Spring		0.0230	0.0195	0.3029	0.1101	0.2188	0.0505	0.9353	0.0163
Summer		0.0528	0.0075	0.2468	0.0921	0.2518	0.0685	0.8769	0.0279
Winter		0.3968	0.4766	0.3148	0.1564	0.5331	0.4268	0.7537	0.2124
Total		0.1000	0.2056	0.2863	0.0950	0.2696	0.1868	0.8850	0.0966
	2011								
Autumn		0.0235	0.0723	0.3017	0.0441	0.2351	0.0839	0.9413	0.0552
Spring		0.0291	0.0238	0.3366	0.1435	0.1973	0.0495	0.9095	0.0448
Summer		0.0804	0.0515	0.2123	0.0563	0.2228	0.0822	0.8846	0.0582
Winter		0.1821	0.3913	0.2925	0.0575	0.4878	0.3039	0.8618	0.1942
Total		0.0624	0.1528	0.2874	0.0971	0.2573	0.1566	0.9063	0.0865
	2012								
Autumn		0.0529	0.0456	0.3375	0.0452	0.2340	0.0691	0.9088	0.0480
Spring		0.0244	0.0367	0.3382	0.1281	0.2057	0.0478	0.9430	0.0542
Summer		0.1005	0.0434	0.2259	0.0662	0.2083	0.0373	0.8445	0.0208
Winter		0.4701	0.4720	0.3858	0.2204	0.6269	0.4147	0.6781	0.2440
Total		0.1264	0.2415	0.3179	0.1263	0.2852	0.2258	0.8638	0.1358
	2013								
Autumn		0.0429	0.0175	0.3259	0.0798	0.2420	0.0793	0.8560	0.0574
Spring		0.0182	0.0187	0.3160	0.1168	0.1683	0.0606	0.9365	0.0389
Summer		0.0170	0.0187	0.2218	0.0824	0.2306	0.0708	0.9240	0.0541
Winter		0.5574	0.4954	0.5706	0.3562	0.7178	0.3904	0.6341	0.2710
Total		0.1244	0.2932	0.3407	0.2067	0.2992	0.2681	0.8657	0.1650
	2014								
Autumn		0.0676	0.0362	0.3823	0.0496	0.2369	0.0866	0.7644	0.1494
Spring		0.0221	0.0205	0.3568	0.1677	0.1965	0.0130	0.9376	0.0361
Summer		0.0139	0.0238	0.1954	0.1014	0.2385	0.0363	0.9537	0.0148
Winter		0.3812	0.4967	0.3156	0.1955	0.4692	0.4601	0.7007	0.1910
Total		0.1212	0.2652	0.3125	0.1414	0.2853	0.2296	0.8391	0.1547
	2015								
Autumn		0.0021	0.0021	0.3123	0.0285	0.2428	0.0314	0.9474	0.0133
Spring		0.0440	0.0386	0.3507	0.1349	0.1673	0.0215	0.9199	0.0360
Summer		0.0314	0.0330	0.2616	0.1234	0.2486	0.1469	0.9278	0.0539
Winter		0.3722	0.4874	0.3429	0.2105	0.4552	0.4718	0.6810	0.2140
Total		0.1124	0.2616	0.3169	0.1250	0.2785	0.2390	0.8690	0.1486
	2016								
Autumn		0.0443	0.0082	0.3238	0.0165	0.2390	0.0273	0.9209	0.0068
Spring		0.0553	0.0437	0.3345	0.1226	0.2279	0.0868	0.8508	0.1155
Summer		0.0755	0.0304	0.2021	0.0764	0.2380	0.0832	0.8876	0.0036
Winter		0.3290	0.5173	0.2690	0.1228	0.4970	0.4370	0.7450	0.2567
Total		0.1260	0.2536	0.2824	0.0980	0.3005	0.2270	0.8511	0.1385
	2017								
Autumn		0.0273	0.0206	0.3193	0.0416	0.2226	0.0298	0.9324	0.0348
Spring		0.0188	0.0182	0.3501	0.1970	0.2208	0.0481	0.9483	0.0277
Summer		0.0494	0.0021	0.3012	0.0305	0.2127	0.0419	0.9225	0.0085
Winter		0.3347	0.5109	0.3765	0.2708	0.4963	0.4366	0.7259	0.2737
Total		0.1075	0.2579	0.3368	0.1476	0.2881	0.2266	0.8823	0.1516

Figure 3 shows a representation of the parameters over time. It can be observed how the values follow a pattern according to the period of the year. The level and AR(1) adjustment values are slightly different from those of the previous analysis. The level values are higher while the AR(1) adjustment is lower. This is because when the trend is removed, possible long-term variations are supplied by the level. The MAPE in winter season worsens while the smoothing parameters increase their variability in autumn. The randomness of the method chosen to analyse variability makes many forecasts done in winter to use a model optimised with data from autumn season. The special events during autumn impact greatly on the winter forecasts.

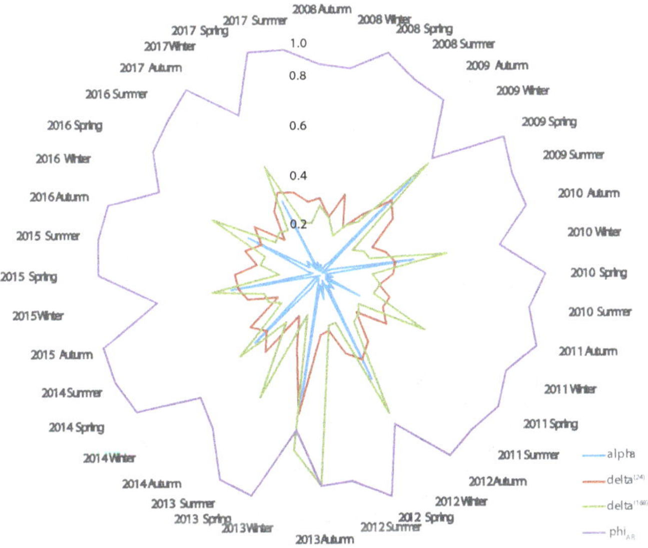

Figure 3. Radar diagram of the average of the parameters for the model $NMC_{24,168}$ depending on the time.

This analysis was completed with a study on the influence of the size of the training dataset on prediction accuracy. The date of 11 July 2016 was set for the 24-hours-ahead forecasts during two weeks, and the size of the training dataset increased from 8 weeks to several years. In this way, it is intended to observe the behaviour of the model with respect to the sample size, following the indications of [41].

Figure ?? presents the MAPE of the 24-hour forecasts according to the size of the sample for the $NMC_{24,168}$ and $AMC_{24,168}$ models in order to compare their behaviour. It can be observed that the $AMC_{24,168}$ model presents higher variability and the model starts to stabilize around the average value 1.7% when a sufficiently large set is used, in particular from 20,000 h. However, the $NMC_{24,168}$ model is much more stable, and with relatively small training sets, in particular, with a time series composed of 5000 h, it maintains accuracy values around 1.6%. The $AMC_{24,168}$ model is more unstable due to the trend equation (Equation (6)) as the electricity demand time series has no clear trend as shown in Figure 1. Using a larger dataset helps the $NMC_{24,168}$ model to stabilize while $AMC_{24,168}$ model depends clearly on the season the dataset starts. Although $NMC_{24,168}$ model shows better stability than $AMC_{24,168}$, the smoothing parameter associated with the trend does not tend to 0 in order to converge to the same model. This is due to the parameters are local optimal, but not global optimal. Thus, the need for an analysis of these parameters and their stability is supported.

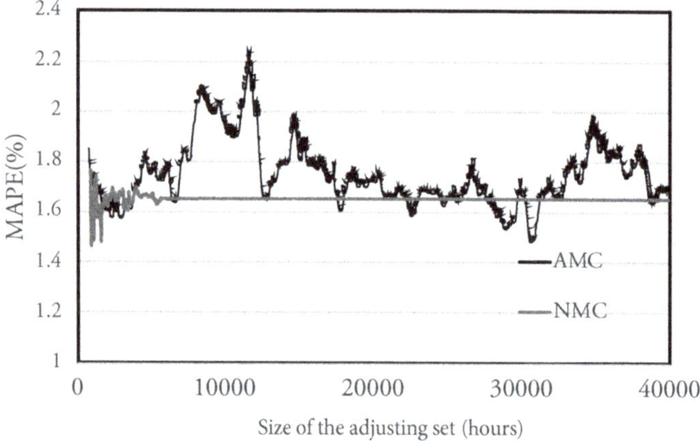

Figure 4. MAPE of the 24-hours ahead forecasts according to the size of the datasets used to obtain the model.

Figure 5 shows the evolution of the value of each parameter according to the size of the data set used for the adjustment of the $NMC_{24,168}$ model. It can be seen how there is a stabilization of the values in an asymptotic way. It is surprising how the values of the parameters associated with the seasonality exchange the weights as the data set grows. It can be noticed some peaks where the values go up. These peaks coincide with the beginning of the series on holidays. In fact, the most intense peaks coincide with the dates of the Immaculate Conception or Christmas. Once again, a stability in the predictions is observed demonstrating a great predictability. This characteristic does not have the same pattern for all models, and depends largely on the values of the seasonal component, always keeping low values for the α smoothing parameter.

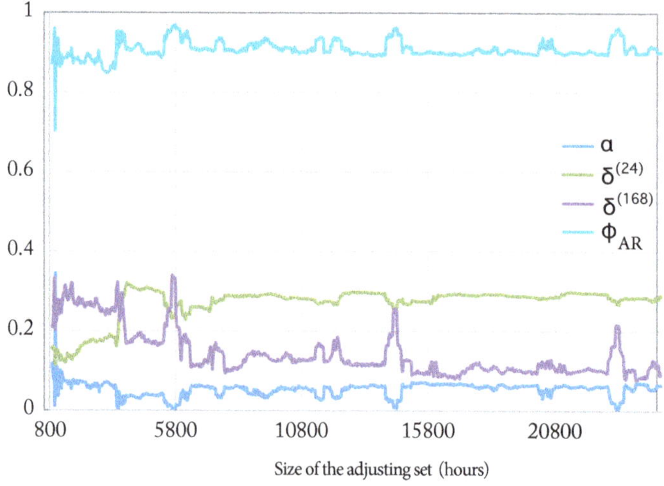

Figure 5. Evolution of parameters versus size of data set.

4.2. Triple Seasonal Models

In this section, an analysis of the parameters of the triple seasonal Holt-Winters models is carried out. Data sets of size 3 years ($53 \times 24 \times 7 \times 3$ h) were used to adjust the models. As in the case of double seasonal models, 24-hour forecasts were made for two weeks. Although all triple seasonal methods depending on the seasonality and trend were analyzed, there are no major differences between the models, and the $AMC_{24,168,8766}$ model turns out to be the easiest to study, as it has 6 parameters, including level, trend, seasonality and fit.

The results obtained are shown in Table 4, where the values of the parameters are organized according to the season and year. In particular, the mean and standard deviation of the smoothing parameters obtained using the 80 sets for adjustment composed of 8 random weeks are shown. It can be noticed that the standard deviation for Autumn of the year 2016 is not defined because it was only a random set of 8 weeks into the Autumn. Contrary to what happens in the double seasonal models, and as seen in the previous section, when using sets of size greater than 20,000 h, the values of the parameters are stabilized.

Table 4. Distribution of smoothing parameters of the $AMC_{24,168,8766}$ model.

	α		γ		$\delta^{(24)}$		$\delta^{(168)}$		$\delta^{(8766)}$		φ_{AR}	
	mean	**Dev.**	**mean**	**Dev.**	**mean**	**Dev.**	**mean**	**Dev.**	**mean**	**Dev.**	**mean**	**Dev.**
2011												
Autumn	0.0008	0.0009	0.0001	0.0000	0.3081	0.0004	0.2296	0.0019	0.0783	0.0306	0.9469	0.0019
Spring	0.0006	0.0008	0.0001	0.0000	0.3137	0.0069	0.2377	0.0183	0.0480	0.0502	0.9463	0.0012
Summer	0.0001	0.0000	0.0001	0.0000	0.3193	0.0094	0.2265	0.0087	0.0718	0.0635	0.9477	0.0014
Winter	0.0001	0.0000	0.0001	0.0000	0.3064	0.0044	0.2144	0.0146	0.1612	0.0038	0.9466	0.0004
Total	0.0003	0.0005	0.0001	0.0000	0.3128	0.0081	0.2267	0.0149	0.0892	0.0604	0.9466	0.0014
2012												
Autumn	0.0001	0.0000	0.0001	0.0000	0.3132	0.0113	0.2446	0.0082	0.0483	0.0682	0.9498	0.0016
Spring	0.0008	0.0009	0.0001	0.0000	0.3032	0.0045	0.2277	0.0018	0.0886	0.0221	0.9438	0.0008
Summer	0.0007	0.0013	0.0001	0.0000	0.3190	0.0090	0.2340	0.0139	0.0684	0.0487	0.9442	0.0017
Winter	0.0004	0.0005	0.0001	0.0000	0.3259	0.0101	0.2404	0.0215	0.0826	0.0386	0.9446	0.0016
Total	0.0005	0.0008	0.0001	0.0000	0.3158	0.0116	0.2344	0.0138	0.0735	0.0405	0.9458	0.0026
2013												
Autumn	0.0001	0.000	0.0001	0.0000	0.3185	0.0079	0.2348	0.0122	0.0802	0.0154	0.9510	0.0003
Spring	0.0001	0.0000	0.0001	0.0000	0.3038	0.0046	0.2280	0.0075	0.0885	0.0328	0.9534	0.0017
Summer	0.0001	0.0000	0.0001	0.0000	0.3147	0.0030	0.2214	0.0060	0.1017	0.0259	0.9515	0.0011
Winter	0.0001	0.0000	0.0001	0.0000	0.3179	0.0133	0.2292	0.0020	0.1481	0.0571	0.9522	0.0014
Total	0.0001	0.0000	0.0001	0.0000	0.3134	0.0090	0.2272	0.0080	0.1064	0.0416	0.9523	0.0014
2014												
Autumn	0.0001	0.0000	0.0001	0.0000	0.3178	0.0100	0.2452	0.0077	0.0467	0.0462	0.9437	0.0005
Spring	0.0001	0.0000	0.0001	0.0000	0.3145	0.0029	0.2251	0.0072	0.0893	0.0126	0.9419	0.0004
Summer	0.0001	0.0000	0.0001	0.0000	0.3235	0.0130	0.2352	0.0094	0.0557	0.0517	0.9429	0.0018
Winter	0.0001	0.0000	0.0001	0.0000	0.3244	0.0032	0.2318	0.0028	0.1003	0.0515	0.9446	0.0022
Total	0.0001	0.0000	0.0001	0.0000	0.3240	0.0084	0.2343	0.0098	0.0731	0.0439	0.9433	0.0016
2015												
Autumn	0.0001	0.0000	0.0001	0.0000	0.2984	0.0095	0.2258	0.0106	0.0983	0.0204	0.9426	0.0014
Spring	0.0001	0.0000	0.0001	0.0000	0.3050	0.0040	0.2279	0.0042	0.0403	0.0147	0.9421	0.0014
Summer	0.0007	0.0010	0.0001	0.0000	0.3134	0.0055	0.2250	0.0127	0.0596	0.0523	0.9411	0.0027
Winter	0.0001	0.0000	0.0001	0.0000	0.3160	0.0068	0.2117	0.0074	0.1377	0.0303	0.9403	0.0003
Total	0.0002	0.0005	0.0001	0.0000	0.3082	0.0092	0.2224	0.0103	0.0840	0.0481	0.9415	0.0017
2016												
Autumn	0.0001	—	0.0001	—	0.3263	—	0.2246	—	0.0399	—	0.9470	—
Spring	0.0001	0.0000	0.00010	0.0000	0.3035	0.0067	0.2262	0.0096	0.0757	0.0273	0.9453	0.0015
Summer	0.0001	0.0000	0.0001	0.0000	0.3112	0.0065	0.2232	0.0055	0.0881	0.0344	0.9457	0.0022
Winter	0.0001	0.0000	0.0001	0.0000	0.3089	0.0040	0.2171	0.0025	0.1301	0.0381	0.9430	0.0008
Total	0.0002	0.0005	0.0001	0.0000	0.3134	0.0097	0.2281	0.0115	0.0862	0.0462	0.9456	0.0038

The value of the $\delta^{(24)}$ parameter stabilizes around 0.3 for any period and with a minimum variability. The value of the $\delta^{(168)}$ parameter stabilizes around 0.2 and the new parameter $\delta^{(8766)}$ around 0.08 and 0.1. These values follow the pattern shown in Figure 5, although there is a transfer between the $\delta^{(168)}$ parameter and the $\delta^{(8766)}$. The introduction of the third seasonality has forced the model to update the intra-weekly seasonality with more recent data over time. On the other hand, values of α stabilize around practically 0, and values of φ_{AR} around 0.95. The existence of so much data makes the AR(1) adjustment practically responsible for adapting the level and makes the level equation redundant. This would imply the elimination of a parameter to be optimized.

In addition, an analysis of the size of the data set and its consequences on the forecasts was carried out in the same way as in the previous case. It was used on the same day, 11 July 2016, where the size of the observed data set was gradually increased and 24-hour forecasts were made over two weeks. The results are shown in Figure 6. An asymptotic evolution towards a MAPE of 2% can be seen. Although the MAPE is triggered and reduced again periodically. The shark fin form responds to an adjusting dataset beginning in the autumn, and finishing in the end of autumn—remember that in autumn the calendar effect was much more important than at other times. One of the main characteristics observed in the triple seasonal models is the variability in predictability, mainly produced by the time of year with more holidays, i.e., autumn.

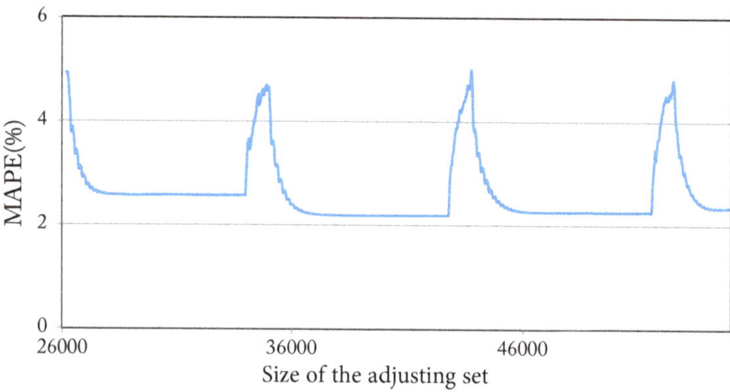

Figure 6. MAPE of the forecasts for 24-hours ahead according to the size of the dataset used to obtain the $AMC_{24,168,8766}$ model.

The selection of 11 July 2016 for the forecasts responds to the necessity to avoid many special events nearby. The nearest one is the 1st May, and the series has enough time to react. As the model has low values for alpha, gamma and δ_{8766}, near to zero, only the intraday and intraweek seasonal components can deal and react against the irregularities of the series, smoothing the model to newer values. The closer the start of the autumn period is, the less reaction time the model has to smooth the irregularities. These irregularities affect forecasts more intensely. When forecasting in other dates, the same analysis provides a similar fin-shaped graph. The only difference is the minimum and maximum of the MAPE, that depends on the date chosen.

5. Discussion of the Results

The objective of this section is the empirical analysis of multiple-seasonal Holt-Winters models applied to hourly electricity demand in Spain.

The literature found that analyses the parameters of the models tries to give a solution to the theoretical stability analysis, with the determination of the concept of predictability. However, its application to Holt-Winters models is not direct. It is necessary to carry out an empirical analysis of the models and their forecasts, and from there to draw conclusions about predictability.

A framework was established consisting of the usual process of adjustment and forecasting using a Spanish hourly electricity demand data set provided by REE. The forecasts obtained are then analyzed.

The double seasonal models with an 8 week adjustment period are shown to be robust with respect to predictions. Two different periods, with different characteristics, were used and 24-hour prediction MAPEs of around 2% to 2.6% were obtained. In the models with the best behaviour, the parameters were analyzed, and a direct relationship was found between variability and high values of the smoothing parameters associated with seasonality, and in the periods when a greater number of holidays occur.

The size of the observed data set influences the stability. The MAPE of the $AMC_{24,168}$ model has a variability of 0.2%, which from about 20,000 h is reduced to 0.1%. The $NMC_{24,168}$ model is much more stable—the series did not show a trend either—achieving this stability in sets longer than 5000 h. As the number of observations increases, the smoothing parameters show a stabilization.

In the case of triple seasonal models, a large number of observations are necessary to adjust the model. Therefore, parameter values are stabilized from the beginning. It can be seen how the predictions get worse when the set of values starts at significant dates in the autumn, a season with many public holidays. Forecasters need to use a dataset avoiding to start in autumn. If no other solution is possible, it is interesting to reduce the dataset size as it is large enough, but not including autumn.

In short, it can be seen that the parameters need a large data set to stabilize, and that the triple seasonal models are not able to improve the double seasonal forecasts due to the calendar effect. As a consequence, it is necessary to develop models that are able to reduce this variability by including the calendar effect in the model.

6. Conclusions

This article analyses the stability of the smoothing parameters in the multiple seasonal Holt-Winters models. This is crucial for the proper appliance of these models to provide accurate and trusted forecasts. Although most of the time, an automatic forecasting algorithm can provide good results, forecasters need to understand the behaviour of these parameters before submitting the forecasts. We use the time series of the hourly Spanish electricity demand.

In this work, we analyze the behaviour of the smoothing parameters of the multiple seasonal Holt-Winters models applied to the series of hourly electricity demand in Spain. There are many variables that affect the parameters within the same time series, including the computation of the initial values, in addition to other factors, such as the calendar or climate conditions. This variability of the parameters is subsequently reflected in the accuracy of the forecasts since they depend greatly on the calculation of the parameters.

The variation of the parameters is analyzed when different seasonal and trend methods are used, as well as different climatic situations of the series. Additionally, it is analyzed how the size of the data set used in order to adjust the model influences on the parameters, and, thus, in the forecasts. It was observed that the seasonal parameters are strongly dependent on the period of the year, making autumn a really difficult period to deal with. This effect is more pronounced in the intra-weekly seasonality than the daily one. However, with the increase in the size of the fitting set, these parameter values stabilise around a value, which depends on the time series. When the size of the set is bigger than 5000 hours, the values are stable. Triple seasonal models have much more stable parameters, although a much larger set of data is necessary to adjust the model. It was found that the main source of variability of the parameters is the calendar effect, which strongly influences the accuracy of the forecasts, not being so much the climatic effect on the series. The accuracy of the forecasts is also stabilised with a larger set of data, it is not necessary to have more than 5000 observations. It is also observed how double seasonal models provide better predictions than triple seasonal ones. The results of the current analysis are limited to being used with Spanish electricity demand. Of course, similar results are expected when faced with load forecasting in other countries or systems. Future works will be addressed toward the development of new models able to include calendar effect in the own model as well as models for the prediction of holidays through the inclusion of discrete seasonalities.

Author Contributions: Conceptualization, Ó.T. and J.C.G.-D.; methodology, J.C.G.-D.; software, Ó.T.; validation, Ó.T., A.T. and J.C.G.-D.; writing–original draft preparation, Ó.T. and J.C.G.-D.; writing–review and editing, A.T.; supervision, A.T. and J.C.G.-D.; project management: A.T. All authors have read and agreed to the published version of the manuscript.

Funding: This research received no external funding.

Acknowledgments: The authors would like to thank the Spanish Ministry of Economy and Competitiveness for the support under project TIN2017-8888209C2-1-R.

Conflicts of Interest: The authors declare no conflict of interest.

References

1. Martínez-Ramos, J.L.; Troncoso, A.; Riquelme-Santos, J.; Gómez-Expósito, A. Short-term hydro-thermal coordination based on interior point nonlinear programming and genetic algorithms. In Proceedings of the IEEE Porto Power Tech Conference, Porto, Portugal, 10–13 September 2001; pp. 1–6.
2. Hobbs, B.F. Analysis of the value for unit commitment of improved load forecasts. *IEEE Trans. Power Syst.* **1999**, *14*, 1342–1348. [CrossRef]
3. Hong, T. Crystal Ball Lessons in Predictive Analytics. *Energybiz* **2015**, *12*, 35–37.
4. Weron, R. *Modeling and Forecasting Electricity Loads and Prices: A Statistical Approach*; John Wiley & Sons: Hoboken, NJ, USA, 2013.
5. Weron, R. Electricity price forecasting: A review of the state-of-the-art with a look into the future. *Int. J. Forecast.* **2014**, *30*, 1030–1081. [CrossRef]
6. Troncoso, A.; Riquelme-Santos, J.M.; Gómez-Expósito, A.; Martínez-Ramos, J.L.; Riquelme, J.C. Electricity Market Price Forecasting Based on Weighted Nearest Neighbors Techniques. *IEEE Trans. Power Syst.* **2007**, *22*, 1294–1301.
7. Chatfield, C.; Mohammad, Y. Holt-Winters Forecasting: Some Practical Issues. *J. R. Stat. Soc. Ser. D (The Statistician)* **1988**, *37*, 129–140. [CrossRef]
8. Gardner, E.S. Exponential smoothing: The state of the art—Part II. *Int. J. Forecast.* **2006**, *22*, 637–666. [CrossRef]
9. Taylor, J.W. Short-term electricity demand forecasting using double seasonal exponential smoothing. *J. Oper. Res. Soc.* **2003**, *54*, 799–805. [CrossRef]
10. Taylor, J.W. Triple seasonal methods for short-term electricity demand forecasting. *Eur. J. Oper. Res.* **2010**, *204*, 139–152. [CrossRef]
11. Taylor, J.W. An evaluation of methods for very short-term load forecasting using minute-by-minute British data. *Int. J. Forecast.* **2008**, *24*, 645–658. [CrossRef]
12. Taylor, J.W.; Espasa, A. Energy forecasting. *Int. J. Forecast.* **2008**, *24*, 561–565. [CrossRef]
13. García-Díaz, J.; Trull, O. Competitive Models for the Spanish Short-Term Electricity Demand Forecasting. In Proceedings of the International Conference on Time Series and Forecasting, Granada, Spain, 27–29 June 2016; pp. 217–231.
14. Trull, O.; García-Díaz, J.C.; Troncoso, A. Initialization Methods for Multiple Seasonal Holt–Winters Forecasting Models. *Mathematics* **2020**, *8*, 268. [CrossRef]
15. Makridakis, S.; Wheelwright, S.; Hyndman, R. *Forecasting: Methods and Applications*; John Willey and Sons: Hoboken, NJ, USA, 1998.
16. Trull, O.; García-Díaz, J.C.; Troncoso, A. Application of Discrete-Interval Moving Seasonalities to Spanish Electricity Demand Forecasting during Easter. *Energies* **2019**, *12*, 1083. [CrossRef]
17. López, M.; Sans, C.; Valero, S.; Senabre, C. Classification of Special Days in Short-Term Load Forecasting: The Spanish Case Study. *Energies* **2019**, *12*, 1253. [CrossRef]
18. Arora, S.; Taylor, J. Short-term forecasting of anomalous load using rule-based triple seasonal methods. *IEEE Trans. Power Syst.* **2013**, *28*, 3235–3242. [CrossRef]
19. David, A.K.; Wen, F. Strategic bidding in competitive electricity markets: a literature survey. In Proceedings of the 2000 Power Engineering Society Summer Meeting (Cat. No. 00CH37134), Seattle, WA, USA, 16–20 July 2000; Volume 4, pp. 2168–2173.
20. Barroso, L.A.; Cavalcanti, T.H.; Giesbertz, P.; Purchala, K. Classification of electricity market models worldwide. In Proceedings of the International Symposium CIGRE/IEEE PES, San Antonio, TX, USA, 5–7 October 2005; pp. 9–16.
21. Roldán-Fernández, J.; Gómez-Quiles, C.; Merre, A.; Burgos-Payán, M.; Riquelme-Santos, J.M. Cross-border energy exchange and renewable premiums: The case of the Iberian system. *Energies* **2018**, *11*, 3277. [CrossRef]

22. Domínguez, E.F.; Bernat, J.X. Restructuring and generation of electrical energy in the Iberian Peninsula. *Energy Policy* **2007**, *35*, 5117–5129. [CrossRef]
23. Cancelo, J.R.; Espasa, A.; Grafe, R. Forecasting the electricity load from one day to one week ahead for the Spanish system operator. *Int. J. Forecast.* **2008**, *24*, 588–602. [CrossRef]
24. López, M.; Valero, S.; Senabre, C. Short-term load forecasting of multiregion systems using mixed effects models. In Proceedings of the 2017 14th International Conference on the European Energy Market (EEM), Dresden, Germany, 6–9 June 2017; pp. 1–5.
25. Talavera-Llames, R.; Pérez-Chacón, R.; Troncoso, A.; Martínez-Álvarez, F. Big data time series forecasting based on nearest neighbours distributed computing with Spark. *Knowl. Based Syst.* **2018**, *161*, 12–25. [CrossRef]
26. Bedi, J.; Toshniwal, D. Deep learning framework to forecast electricity demand. *Appl. Energy* **2019**, *238*, 1312–1326. [CrossRef]
27. Torres, J.F.; Troncoso, A.; Koprinska, I.; Wang, Z.; Martínez-Álvarez, F. Big data solar power forecasting based on deep learning and multiple data sources. *Appl. Energy* **2019**, *238*, 1312–1326. [CrossRef]
28. Yang, Y.; Hong, W.; Li, S. Deep ensemble learning based probabilistic load forecasting in smart grids. *Energy* **2019**, *189*, 116324. [CrossRef]
29. Galicia, A.; Talavera-Llames, R.; Troncoso, A.; Koprinska, I.; Martínez-Álvarez, F. Multi-step forecasting for big data time series based on ensemble learning. *Knowl. Based Syst.* **2019**, *163*, 830–841. [CrossRef]
30. Jiang, W.; Wu, X.; Gong, Y.; Yu, W.; Zhong, X. Holt–Winters smoothing enhanced by fruit fly optimization algorithm to forecast monthly electricity consumption. *Energy* **2020**, *193*, 116779. [CrossRef]
31. Holt, C.C. Forecasting seasonals and trends by exponentially weighted moving averages. *Int. J. Forecast.* **2004**, *20*, 5–10. [CrossRef]
32. Bowerman, B.L.and O'Connell, R.; Koehler, A. *Forecasting, Time Series, and Regression: An Applied Approach*, 4th ed.; Thomson Brooks/Cole: Belmont, CA, USA, 2005.
33. Makridakis, S.; Hibon, M. Exponential smoothing: The effect of initial values and loss functions on postsample forecasting accuracy. *Int. J. Forecast.* **1991**, *7*, 317–330. [CrossRef]
34. Archibald, B.C. Parameter space of the Holt–Winters' model. *Int. J. Forecast.* **1990**, *6*, 199–210. [CrossRef]
35. Lawton, R. How should additive Holt–Winters estimates be corrected? *Int. J. Forecast.* **1998**, *14*, 393–403. [CrossRef]
36. Barnett, S.; Cameron, R.G. *Introduction to Mathematical Control Theory*, 2nd ed.; Oxford University Press: Oxford, UK, 1985.
37. Harvey, A.C. *Forecasting, Structural Time Series Models and the Kalman Filter*; Cambridge University Press: Cambridge, UK, 1990.
38. Hyndman, R.J.; Akram, M.; Archibald, B.C. *Invertibility Conditions for Exponential Smoothing Models*; Monash University, Department of Econometrics and Business Statistics: Monash, Australia, 2003.
39. Hyndman, R.J.; Akram, M.; Archibald, B.C. The admissible parameter space for exponential smoothing models. *Ann. Inst. Stat. Math.* **2008**, *60*, 407–426. [CrossRef]
40. Osman, A.F.; King, M.L. Stability and Forecastability Characteristics of Exponential Smoothing with Regressors Methods. In Proceedings of the Regional Conference on Science, Technology and Social Sciences (RCSTSS 2016), Penang, Malaysia, 4–6 December 2016; pp. 1029–1038.
41. Hyndman, R.J.; Kostenko, A.V. Minimum sample size requirements for seasonal forecasting models. *Foresight* **2007**, *6*, 12–15.
42. Bermúdez, J. Exponential smoothing with covariates applied to electricity demand forecast. *Eur. J. Ind. Eng.* **2013**, *7*, 333–349. [CrossRef]
43. Troncoso, A.; Riquelme-Santos, J.M.; Riquelme, J.; Gómez-Expósito, A.; Martínez-Ramos, J.L. Time-Series Prediction: Application to the Short-Term Electric Energy Demand. *Lecture Notes Comput. Sci.* **2004**, *3040*, 577–86.
44. Rana, M.; Koprinska, I. Forecasting electricity load with advanced wavelet neural networks. *Neurocomputing* **2016**, *182*. [CrossRef]
45. Nelder, J.A.; Mead, R. A Simplex Method for Function Minimization. *Comput. J.* **1965**, *7*, 308–313. [CrossRef]

46. Hyndman, R.J.; Koehler, A.B. Another look at measures of forecast accuracy. *Int. J. Forecast.* **2006**, *22*, 679–688. [CrossRef]
47. Tofallis, C. A better measure of relative prediction accuracy for model selection and model estimation. *J. Oper. Res. Soc.* **2015**, *66*, 1352–1362. [CrossRef]

Article

Temporal Convolutional Networks Applied to Energy-Related Time Series Forecasting

Pedro Lara-Benítez *,†, Manuel Carranza-García †, José M. Luna-Romera and José C. Riquelme

Division of Computer Science, University of Sevilla, ES-41012 Seville, Spain; mcarranzag@us.es (M.C.-G.); jmluna@us.es (J.M.L.-R.); riquelme@us.es (J.C.R.)

* Correspondence: plbenitez@us.es

† These authors contributed equally to this work.

Received: 4 March 2020; Accepted: 24 March 2020; Published: 28 March 2020

Featured Application: Energy demand forecasting to improve power generation management.

Abstract: Modern energy systems collect high volumes of data that can provide valuable information about energy consumption. Electric companies can now use historical data to make informed decisions on energy production by forecasting the expected demand. Many deep learning models have been proposed to deal with these types of time series forecasting problems. Deep neural networks, such as recurrent or convolutional, can automatically capture complex patterns in time series data and provide accurate predictions. In particular, Temporal Convolutional Networks (TCN) are a specialised architecture that has advantages over recurrent networks for forecasting tasks. TCNs are able to extract long-term patterns using dilated causal convolutions and residual blocks, and can also be more efficient in terms of computation time. In this work, we propose a TCN-based deep learning model to improve the predictive performance in energy demand forecasting. Two energy-related time series with data from Spain have been studied: the national electric demand and the power demand at charging stations for electric vehicles. An extensive experimental study has been conducted, involving more than 1900 models with different architectures and parametrisations. The TCN proposal outperforms the forecasting accuracy of Long Short-Term Memory (LSTM) recurrent networks, which are considered the state-of-the-art in the field.

Keywords: deep learning; energy demand; temporal convolutional network; time series forecasting

1. Introduction

Forecasting electricity demand is currently amongst the most important challenges for the industries. Due to the increasingly high level of electricity consumption, electrical companies need to efficiently manage the production of energy. Sustainable production plans are required to meet demands and account for important challenges of this century such as global warming and the energy crisis. Smart meters now provide useful data that can help to understand consumption patterns and monitor power demand more efficiently. Data mining techniques can use this information to learn from historical past data and predict the expected demand to make decisions accordingly. Obtaining accurate forecasts can be essential for the future electricity market considering the increasing penetration of renewable energies. However, forecasting power demand is a complex task that involves many factors and requires sophisticated machine learning models to produce high-quality predictions.

Statistical-based models, such as the Box–Jenkins model called ARIMA, were for many years the state-of-the-art for electricity time series forecasting [1,2]. However, machine learning models have proven to provide better performance for problems of this domain. Artificial neural networks (ANNs) [3], support vector machines (SVMs) [4,5], and regression trees [6] have been applied successfully for diverse power demand prediction tasks. More recently, deep learning (DL) has emerged as a very powerful approach for time series forecasting. DL models are especially suitable for big-data temporal sequences due to their capacity to extract complex patterns automatically without feature extraction preprocessing steps [7]. As an evolution from simple ANNs, deep, fully connected networks have been applied for load forecasting problems [8]. However, fully connected networks are unable to capture the temporal dependencies of a time series. Consequently, more specialised DL models such as recurrent neural networks (RNNs) and convolutional neural networks (CNNs) started to gain importance in the time series forecasting field. These networks can efficiently encode the underlying patterns of time series by transforming the temporal problem into a spatial architecture [9].

In the recent literature, a significant number of studies presenting results of the application of RNNs to energy-related time series forecasting can be found [10,11]. Among all existing RNN architectures, long short-term memory (LSTM) networks have been the most popular due to their capacity to solve problems of previous RNN such as gradient explosion and vanishing gradient [12]. It has been considered a standard forecasting model for several tasks such as traffic prediction [13], solar power forecasting [14], financial market predictions [15], and electricity price prediction [16]. Although CNNs were originally designed for computer vision tasks, they are also suitable for time series data since they can extract high-level features from data with a grid topology. Despite the popularity of RNNs, several works using convolutional networks can be found. In both [17,18], the authors proposed CNN models for short-term load forecasting that provides comparable results to LSTM models. Other works have been able to build deep convolutional networks that can outperform LSTM networks for electricity demand [19] and solar power data problems [20]. Furthermore, in all these works, the CNN models proved to be more suitable for real-time applications given their faster training and testing execution time. The properties of local connectivity and parameter sharing of convolutional networks reduce the number of trainable parameters compared to RNNs, hence they can be trained more efficiently. There have also been proposals using hybrid models that combine convolutional and LSTM layers. In [21], the output feature maps of a CNN are fed to a RNN that provides the prediction. Other approaches consider combining the features extracted in parallel from a CNN and a LSTM to improve the forecasting using electricity demand data [22] or financial data [23]. These ensemble proposals can enhance the predictive performance by fusing the long-term patterns captured by the LSTM and the local trend features obtained with the CNN.

More recently, a specialised CNN architecture known as temporal convolutional networks (TCN) has acquired popularity due to their suitability to deal with time series data. TCNs were first proposed in [24], in which they were compared to several RNNs over sequence modelling tasks. TCNs use causal dilated causal convolution in order to be able to capture longer-term dependencies and prevent information loss. Furthermore, they present other advantages over RNNs such as lower memory requirements, parallel processing of long sequences as opposed to the sequential approach of RNNs, and a more stable training scheme. Several works have already successfully used TCNs for time series forecasting tasks: the original architecture using stacked dilated convolutions was proposed in [25] to improve the performance of LSTM networks for financial domain problems; Ref. [26] designed a deep TCN for multiple related time series with an encoder–decoder scheme, evaluating over data from the sales domain; the study in [27] proposed a multivariate time series forecasting model for meteorological data, which outperformed several popular deep learning models. However, to the best of our knowledge, the potential of TCNs has not yet been explored for univariate time series forecasting problems related to electricity demand data.

In this work, we study the applicability and performance of TCNs for multistep time series forecasting over two energy-related datasets. With the first dataset, we build a deep learning model to forecast the electricity demand in Spain based on the historical consumption data over five years. In the second dataset, the problem is to forecast the expected energy consumption of charging stations for electric vehicles in Spain. Our aim in this study is to present a deep learning model that uses a TCN to obtain high accuracy on time series forecasting. We present the results obtained with several TCN architectures and perform an extensive comparison with different LSTM models, which has been so far the most extended approach for these types of problems. In the experimental study, we carry out an extensive parameter search process which involves 1998 different network architectures.

In summary, the main scientific contributions of this paper can be condensed as follows:

- A temporal convolutional neural network model to achieve high accuracy in forecasting over energy demand time series;
- A thorough experimental study, comparing the performance of temporal convolutional with long short-term memory networks for time series forecasting.

The rest of the paper is organised as follows: Section 2 describes the materials used, the methodology, and the experiments carried out; in Section 3, the experimental results obtained are reported and discussed; Section 4 presents the conclusions and future work.

2. Materials and Methods

In this section, we present the datasets selected for the study, the methodology to perform time series forecasting using deep learning models, and the details of the experimental study carried out.

2.1. Datasets

In this subsection, we present the two energy-related datasets selected for the study.

2.1.1. Electric Demand in Spain

The first dataset used in the experimental study covers the national electrical energy demand in Spain ranging from 2014-01-02T00:00 to 2019-11-01T23:50. During this period, the consumption was measured every ten minutes which makes a time series with a total length of 306,721 measurements. The data was provided by Red Eléctrica de España (the Spanish public grid) and is publicly available at [28]. The dataset was divided into two sets: the training set contains 245,376 samples corresponding to the period from 2014-01-02T00:00 to 2018-09-02T00:50; and the test set contains 61,343 samples comprising the period from 2018-09-02T01:00 to 2019-11-01T23:40. As it has been done in previous studies using this data [29], we defined the forecasting horizon to be 4 hours, which involves a prediction of 24 time-steps.

Figure 1 plots the electric demand data at different scales. As can be seen, the time series presents both weekly and daily seasonality. In general, the demand is higher at weekdays than at weekends and suffers a severe drop during night-time every day.

2.1.2. Electric Vehicles Power Consumption

The second dataset gathers information about power consumption in charging stations for electric vehicles (EV) in Spain. This data was also obtained from Red Eléctrica de España and can be found at [30]. In the near future, governments will have to build infrastructures that can fulfil the demands of the increasing EV fleet. Given their limited autonomy, the prediction of EV consumption seems crucial to efficiently manage the power supply. In this article, we followed the same steps to generate the EV demand time series as in [31]. The data was collected hourly and ranges from 2015-03-02T00:00 to 2016-05-31T23:00. For each geographical area, we obtained a single value of power consumption for every hour. Later, in order to obtain a single time series, the different zones were aggregated giving the total energy consumption of the Spanish EV. The forecasting horizon was set to be 48 h, which involves a prediction of 48 time-steps. The dataset is divided into two sets: the training set contains 8759 samples corresponding to the period from 2015-03-02T00:00 to 2016-02-29T23:00; and the test set contains 2207 samples comprising the period from 2016-03-01T00:00 to 2016-05-31T23:00.

As can be seen in Figure 2, the time series presents weekly and daily patterns. Since electric vehicles are most commonly charged at night, the time series presents peak values of demand during the first six hours of each day. Only Sundays and Mondays present a slightly different pattern, as it is displayed in Figure 2c.

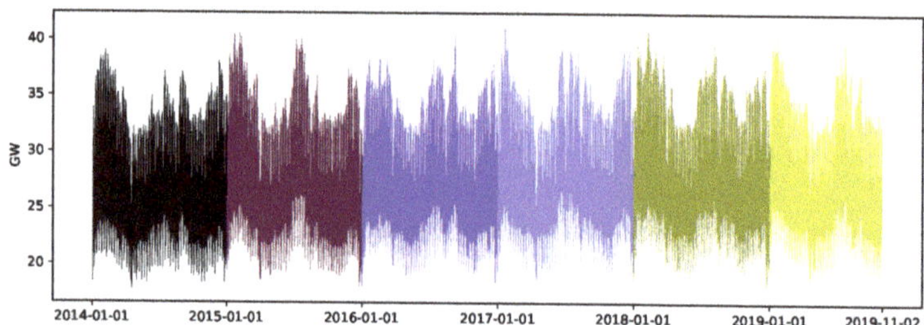

(**a**) Complete time series showing the evolution of the electric demand in Spain from 2014 to 2019.

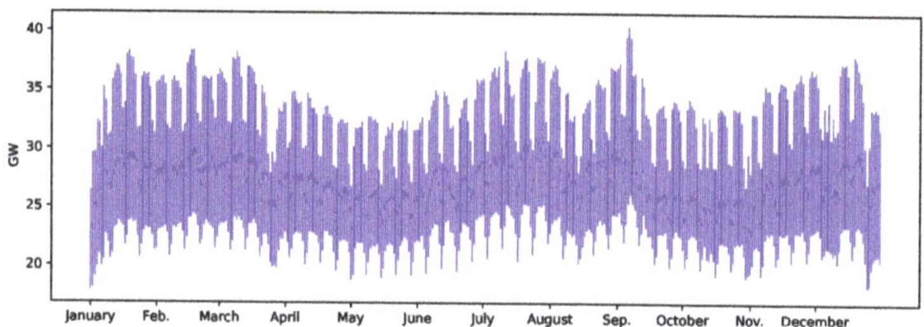

(**b**) Evolution of the electric demand in Spain during 2016.

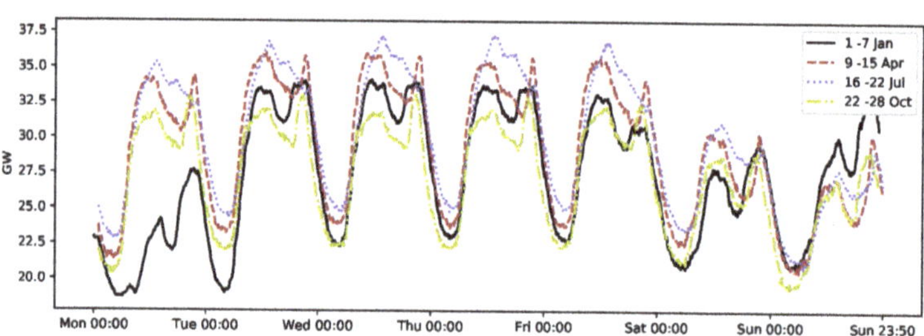

(**c**) Evolution of the electric demand during four different weeks of 2018.

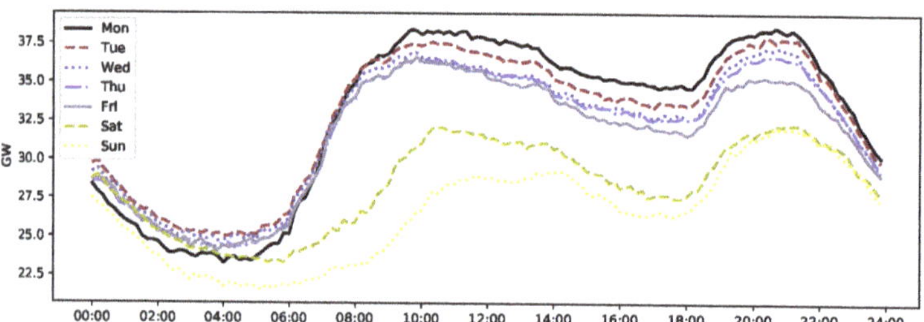

(**d**) Evolution of the electric demand within each day of the first week of February 2019.

Figure 1. Line plots illustrating the electric demand time series data at different scales.

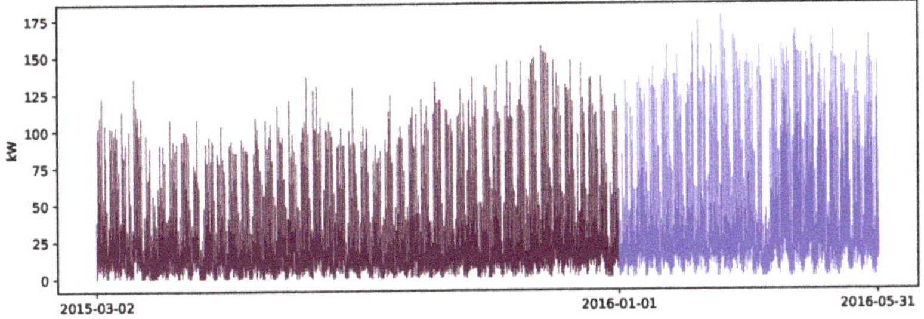

(a) Complete time series showing the evolution of electric vehicle (EV) power consumption from March 2015 to end of May 2016.

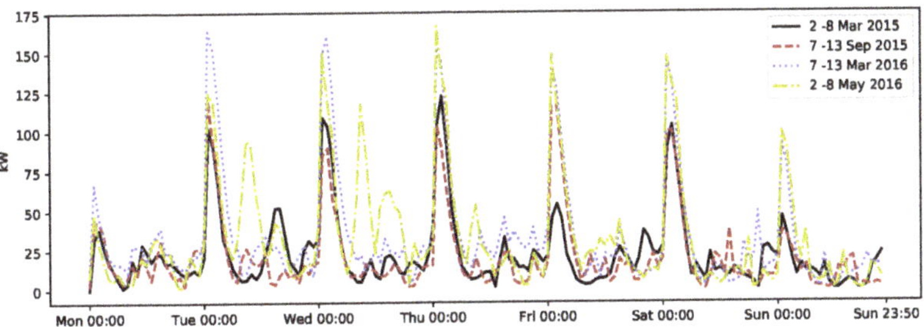

(b) Evolution of the EV power consumption during four different weeks.

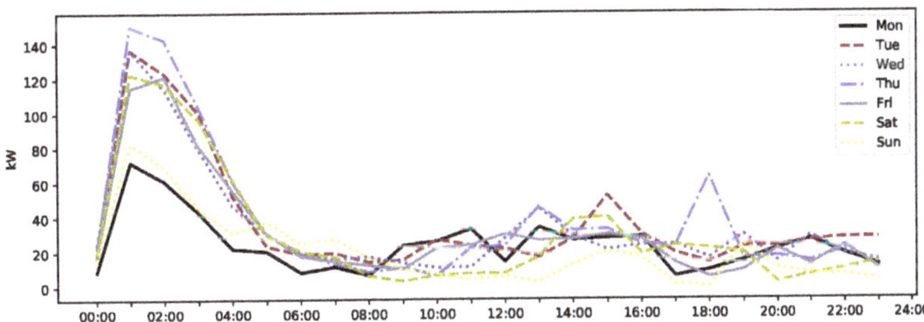

(c) Evolution of the EV power consumption within each day of the first week of February 2016.

Figure 2. Line plots illustrating the electric vehicle power consumption data at different scales.

2.2. Methodology

In this subsection, we describe the required data preprocessing steps and the fundamental concepts behind temporal convolutional networks.

2.2.1. Data Preprocessing

In order to train a deep learning model that can predict several time-steps, a preprocessing stage is needed to transform the original time series data. First, we perform min-max normalisation to the entire sequence to scale the values between 0 and 1, which helps to improve the convergence of deep networks. Secondly, we transform the sequence into instances that can be used to feed the network. There exist several strategies to deal with multistep forecasting problems [32]: the recursive strategy, which performs one-step predictions and feeds the result as the last input for the next prediction; the direct strategy, which builds one model for each time step; and the multi-output approach, which outputs the complete forecasting horizon vector using just one model. As suggested in recent forecasting studies that use neural networks [33,34], in this work, we adopt the MIMO strategy (Multi-Input Multi-Output) which belongs to the last category. Instead of forecasting each time-step independently, the MIMO approach can model the dependencies between the predicted values since it outputs the complete forecasting window. Furthermore, this strategy avoids the accumulated errors over predictions that appear in the recursive strategy.

Following this approach, a moving window scheme is used to create the input–output pairs that will be fed to the neural network. All deep learning models used in this study accept a fixed-length window as input and have an output dense layer with as many neurons as the forecasting horizon defined for each problem (24 for electricity demand and 48 for electric vehicle demand). Figure 3 illustrates the process of applying the moving window over the complete time series. As can be seen, the window slides and obtains an input–output instance at each position. While the output window size is defined by the problem, the input window size has to be decided. The optimal value can be different depending on the data, the designed model, and the forecasting horizon. In our study, we have experimented with three different sizes for the input window of each problem. The values have been carefully selected, considering the characteristics and seasonality of the datasets. For the electricity demand, we evaluate using 144, 168, and 268 time-steps as input window (which corresponds to 24, 28, and 48 h, respectively). For the power demand of electric vehicles, we consider 168, 336, and 672 time-steps as input window (which corresponds to 7, 14, and 28 days, respectively).

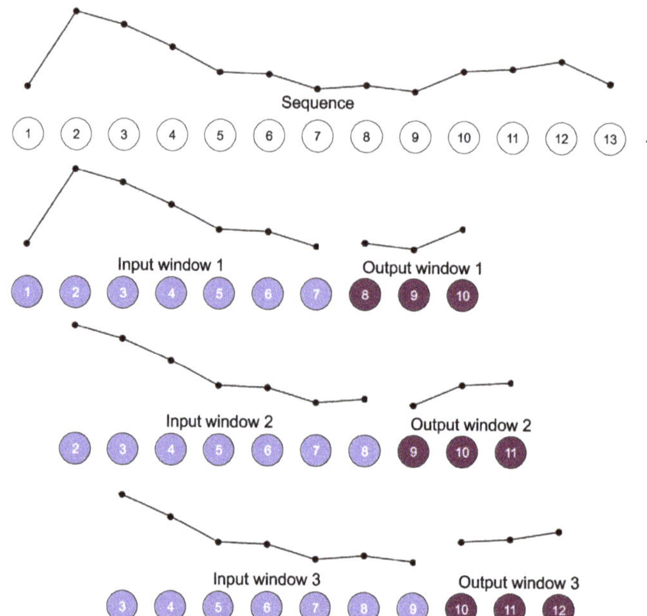

Figure 3. Moving window procedure that obtains the input–output instances. In this example, the input and output windows have lengths of 7 and 3, respectively.

2.2.2. Temporal Convolutional Neural Network

TCNs are a type of convolutional neural network with a specific design that makes them suitable for handling time series. TCNs satisfy two main principles: the network's output has the same length as the input sequence (similarly to LSTM networks); and they prevent leakage of information from future to the past by using causal convolutions [24]. Causal convolution differs from standard convolution in the fact that the convolutional operation performed to obtain the output at time t does not take future values as inputs. This implies that, using a kernel size k, the output O_t is obtained using the values of $X_{t-(k-1)}, X_{t-(k-2)}, \ldots, X_{t-1}, X_t$ (Figure 4). Zero-padding of length $k-1$ is used at every layer to maintain the same length as the input sequence.

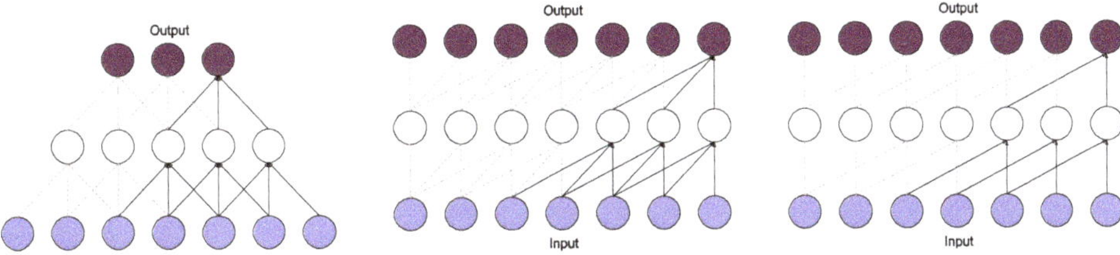

(**a**) Standard convolution block with two layers with kernel size 3.

(**b**) Causal convolution block with two layers with kernel size 3.

(**c**) Dilated causal convolution block with two layers with kernel size 2, dilation rate 2.

Figure 4. Differences between (**a**) standard convolutional network, (**b**) causal convolutional network, and (**c**) dilated causal convolutional network.

Furthermore, with the aim of capturing longer-term patterns, TCNs use one-dimensional dilated convolutions. This convolution increases the receptive field of the network without using pooling operations, hence there is no loss of resolution [35]. Dilation consists of skipping d values between the inputs of the convolutional operation, as can be seen in Figure 4c. The complete dilated causal convolution operation over consecutive layers can be formulated as follows [36]:

$$x_l^t = g\left(\sum_{k=0}^{K-1} w_l^k \, x_{(l-1)}^{(t-(k\times d))} + b_l\right), \tag{1}$$

where x_l^t is the output of the neuron at position (t) in the l-th layer; K is the width of the convolutional kernel; w_l^k stands for the weight of position (k); d is the dilation factor of the convolution; and b_l is the bias term. Rectified Linear Units (ReLU) layers are used as activation function ($g(x) = max(0,x)$) [37]. Another common approach to further increase the network's receptive field is to concatenate several TCN blocks, as can be seen in Figure 5 [38]. However, this leads to deeper architectures with many more parameters which complicates the learning procedure. For this reason, a residual connection is added to the output of each TCN block. Residual connections were proposed by [39] in order to improve performance in very deep architectures, and consist of adding the input of a TCN block to its output ($o = g(x + F(x))$).

All these characteristics make TCNs a very suitable deep learning architecture for complex time series problems. The main advantage of TCNs is that, similarly to RNNs, they can handle variable-length inputs by sliding the one-dimensional causal convolutional kernel. Furthermore, TCNs are more memory efficient than recurrent networks due to the shared convolution architecture which allows them to process long sequences in parallel. In RNNs, the input sequences are processed sequentially, which results in higher computation time. Moreover, TCNs are trained with the standard backpropagation algorithm, hence avoiding the gradient problems of the backpropagation-through-time algorithm (BPTT) used in RNN [40].

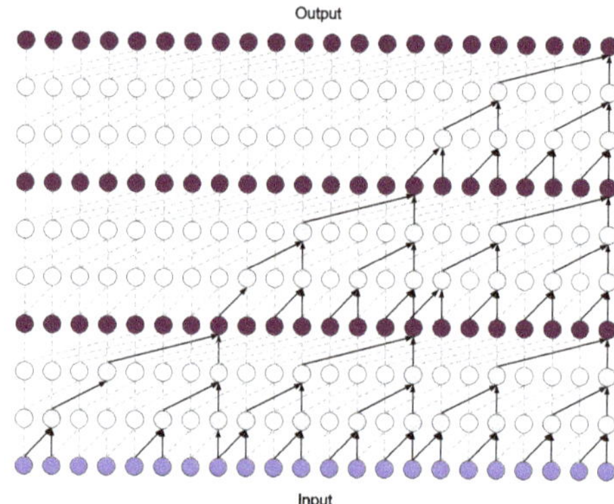

Figure 5. Temporal Convolutional Networks (TCN) model with 3 stacked blocks. Each block has 3 convolutional layers with kernel size 2 and dilations [1, 2, 4].

2.3. Experimental Study

In this subsection, we present the design of the experimental study carried out over the two energy-related datasets. Furthermore, we also describe the details of the parameter search process for each model architecture.

2.3.1. Models

The aim of this work is to build a TCN-based deep learning model to improve the performance in energy demand forecasting tasks, in terms of both accuracy and efficiency. In order to compare the effectiveness of TCNs for this problem, we also evaluate the performance of recurrent LSTM networks—that have so far been considered the state-of-the-art in forecasting. However, given the high complexity of deep learning models, finding optimal values for the hyperparameters of these networks is a very challenging task. Therefore, we have conducted an extensive experimental study that involves more than 1900 combinations of parameters that build different convolutional and recurrent architectures. An important hyperparameter that is common to both types of architectures is the size of the input window. The possible values for past history were defined in Section 2.1 and depend on the data characteristics and seasonality. Additionally, we have searched for the best values of several parameters that are specific for TCN or LSTM architectures. In the case of TCNs, we have experimented with a different number of filters and stacked residual blocks, kernel sizes, and dilations factors. In the case of LSTMs, we have experimented with a different number of stacked layers and units. Furthermore, we have also studied the effect of training parameters in the performance of all models, such as the batch size and the number of epochs. The Adam optimiser has been selected for training the models, which has an adaptive learning rate that can improve the convergence speed of deep networks [41]. The mean absolute error (MAE) has been used as the loss function for all experiments.

Table 1 displays all TCN architecture configurations that have been tested over both datasets. The parameter search process has been designed considering the receptive field of neurons inside the network, that can be calculated as follows: $(receptive\ field = no.\ stacked\ blocks \times kernel\ size \times last\ dilation\ factor)$. Depending on the length of the past history window, we carefully select possible values for kernel size, stacked blocks, and dilations so that the receptive field covers the whole input sequence. For instance, if the number of stacked block increases, less dilated convolutional layers are needed, as can be seen in Table 1a,b. All these architectures are then tested with all combinations of parameters displayed in Table 1c that are common for both datasets (number of convolutional filters, epochs, and batch size). Overall, 756 (28 models from Figure 7a $\times$ 3 numbers of filters $\times$ 3 batch size $\times$

3 number of epochs from Table 1c) experiments with different configurations using TCNs have been conducted for the electric demand dataset, and 513 (12 from Figure 7b × 3×3×3 from Table 1c) for the EV power consumption data.

Table 1. Architecture configuration of all TCN models for each dataset and common parameter search.

a TCN architectures depending on the past history for electric demand data.

Past History	TCN Model Architecture		
	Kernel Size	No. Blocks	Dilations
144	2	3	[1, 3, 6, 12, 24]
	3	1	[1, 3, 6, 12, 24, 48]
		2	[1, 3, 6, 12, 24]
		3	[1, 2, 4, 8, 16]
		4	[1, 3, 6, 12]
	4	3	[1, 3, 6, 12]
		4	[1, 3, 9]
	6	1	[1, 3, 6, 12, 24]
		2	[1, 3, 6, 12]
		3	[1, 2, 4, 8]
		4	[1, 3, 6]
168	2	3	[1, 5, 7, 14, 28]
	3	1	[1, 5, 7, 14, 28, 56]
		2	[1, 5, 7, 14, 28]
		4	[1, 5, 7, 14]
	4	3	[1, 5, 7, 14]
	6	1	[1, 5, 7, 14, 28]
		2	[1, 5, 7, 14]
		4	[1, 5, 7]
288	2	3	[1, 3, 6, 12, 24, 48]
	3	2	[1, 3, 6, 12, 24, 48]
		3	[1, 2, 4, 8, 16, 32]
		4	[1, 3, 6, 12, 24]
	4	3	[1, 3, 6, 12, 24]
	6	1	[1, 3, 6, 12, 24, 48]
		2	[1, 3, 6, 12, 24]
		3	[1, 2, 4, 8, 16]
		4	[1, 3, 6, 12]

b TCN architectures depending on the past history for EV power consumption data.

Past History	TCN Model Architecture		
	Kernel Size	No. Blocks	Dilations
168	2	3	[1, 5, 7, 14, 28]
	3	1	[1, 5, 7, 14, 28, 56]
		2	[1, 5, 7, 14, 28]
		3	[1, 5, 7, 14]
	4	3	[1, 5, 7, 14]
	6	1	[1, 5, 7, 14, 28]
		2	[1, 5, 7, 14]
		4	[1, 5, 7]
336	2	3	[1, 5, 7, 14, 28, 56]
	3	2	[1, 5, 7, 14, 28, 56]
		4	[1, 5, 7, 14, 28]
	4	3	[1, 5, 7, 14, 28]
	6	1	[1, 5, 7, 14, 28, 56]
		2	[1, 5, 7, 14, 28]
		4	[1, 5, 7, 14]
672	3	4	[1, 5, 7, 14, 28, 56]
	4	3	[1, 5, 7, 14, 28, 56]
	6	2	[1, 5, 7, 14, 28, 56]
		4	[1, 5, 7, 14, 28]

c Parameter search for TCN models.

TCN Parameters	
No. of filters	{32, 64, 128}
Batch size	{64, 128, 255}
No. of epochs	{25, 50, 100}

Table 2 presents the parameters that have been studied for LSTM networks. Given the sequential processing nature of these networks, they can effectively cover the complete input sequence and capture long-term dependencies. Therefore, the parameter search, in this case, is based on trying different combinations of LSTM units that can process sequences and feed the output to subsequent stacked layers. We also consider the same possible values as above for the input window length. A total of 243 (3 past history × 3×3×3 from Table 2) experiments with different LSTM models have been carried out for each dataset.

Table 2. Parameter search for LSTM models for both datasets.

LSTM Parameters		
No. stacked layers		{1, 2, 3}
LSTM Units		{32, 64, 128}
Training parameters	**Batch size**	{64, 128, 256}
	No. of epochs	{25, 50, 100}

2.3.2. Evaluation Metric

For evaluating the predictive performance of all models we use the weighted absolute percentage error (WAPE). This metric has been suggested by recent studies dealing with energy demand data [31]. Electric industries are interested in knowing the deviation in watts for better load generation planning. Therefore, WAPE is very suitable for this context since it provides absolute error values. WAPE can be defined as follows:

$$WAPE(y,o) = \frac{MAE(y,o)}{mean(y)} = \frac{mean(|y-o|)}{mean(y)}, \quad (2)$$

where y and o are two vectors with the real and predicted values, respectively, that have a length equal to the forecasting horizon.

3. Results and Discussion

This section reports and discusses the results obtained from the experiments carried out with the different model architectures presented in the previous section. For all tests, we have used a computer with an Intel Core i7-770K CPU and a NVIDIA GeForce GTX 1080 8GB GPU. The source code and the complete experimental results report can be found at [42].

3.1. Forecasting Accuracy

Figure 6 shows a comparison between the overall performance of TCN and LSTM models. It presents the distribution of the results obtained for each dataset with all architectures depending on the past history window length. In general, it can be seen that TCN models achieve a better predictive accuracy compared to LSTM networks. In almost all cases, groups of different TCN architectures with the same input window have a smaller deviation. This implies that TCN models are less sensitive to the hyperparameter selection as long as the past history remains fixed. The most robust performance is given by TCN architectures using input windows of 288 for the electric demand data and 168 for the EV power demand data. Only in the case of the EV power consumption using a very long past history (672 h) do TCN models struggle to get a stable performance. Using longer input windows implies more trainable parameters and can complicate the learning procedure of very deep convolutional networks. With respect to the LSTM models, the worst results are obtained when using the largest history size. This suggests that the proposed recurrent networks are not able to efficiently process long input sequences, and can better capture temporal dependencies using smaller windows. The difference in performance between TCN and LSTM is higher in the electric demand dataset, which is also the longest time series. This indicates that the acquisition of high volumes of data is a fundamental step in order to obtain robust TCN-based deep learning models.

Table 3 presents the TCN and LSTM architecture that obtained the best WAPE result for each past history value. The highest accuracy for both datasets has been obtained with very similar TCN architectures. In the case of the electric demand data, the best result (0.0093 WAPE) has been obtained using 48 h (288 time-steps) as past history. This TCN architecture, which is represented in Figure 7a, has two residual blocks with five convolutional layers of kernel size 6, 128 filters, and dilations of 1, 3, 6, 12, and 24. This model has been trained over 50 epochs with a batch size of 128 instances. In contrast, the best LSTM model (0.0105 WAPE) for this dataset uses the smallest possible input window, which is 24 h. This LSTM model consists of 2 stacked layers with 128 units, and was trained over 50 epochs using 64 as batch size. In general, for this dataset, it can be seen that the best results have been provided by stacking two or three layers that use the greatest amount of filters or units (128). For TCNs, the highest kernel size (6) has proved to be the most effective for capturing local trend patterns in this long time series with daily and weekly seasonality. Furthermore, TCN blocks needed at least four convolutional layers with increasing dilation factors to achieve the highest accuracy. Both types of network achieve better results when training for 50 epochs, which suggest that an excessive number of iterations may cause overfitting issues. Concerning the batch size, the optimal value is 128 for almost all cases. Deep networks lose generalisation capacity when trained using large batches since they often converge to sharp minimisers [43], hence choosing smaller values can be beneficial.

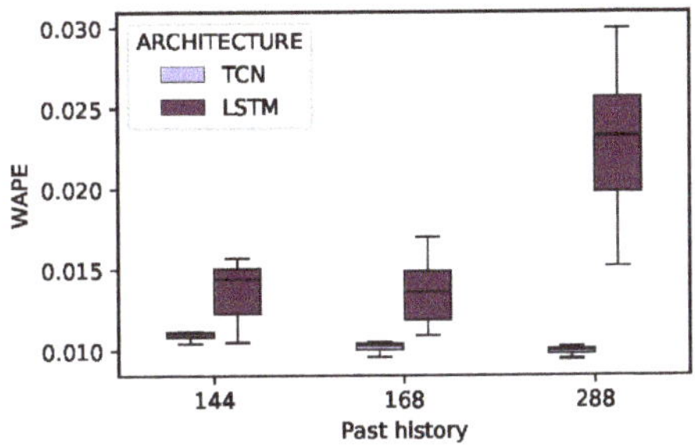

(a) Results for the electric demand data.

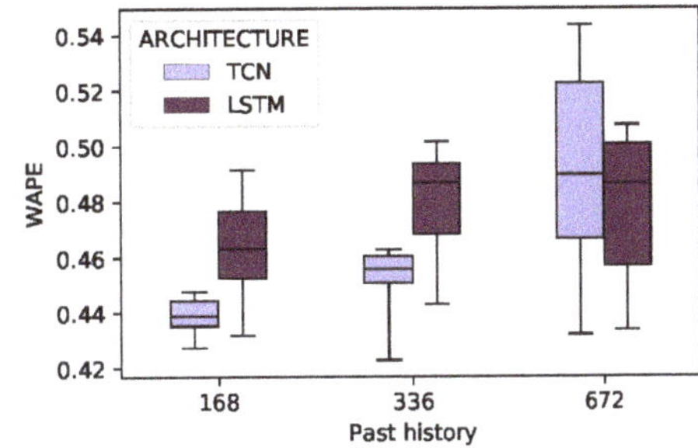

(b) Results for the EV power consumption data.

Figure 6. Distribution of weighted absolute percentage error (WAPE) results for all architectures depending on the past history window.

Table 3. WAPE results of the best TCN and Long Short-Term Memory (LSTM) models for each possible value of input window.

a Best results for electric demand data.

Past History		Model Description								WAPE
Time-Steps	Hours	#	Type	Blocks	Filters	Kernel	Dilations	Batch Size	Epochs	
144	24	1	LSTM	3	128			128	50	0.0105
		2	TCN	2	128	6	[1, 3, 6, 12]	256	50	0.0104
168	28	3	LSTM	3	128			128	50	0.0109
		7	TCN	2	128	6	[1, 5, 7, 14]	128	50	0.0096
288	48	5	LSTM	3	128			128	50	0.0151
		6	TCN	2	128	6	[1, 3, 6, 12, 24]	128	50	**0.0093**

b Best results for the EV power consumption data.

Past History		Model Description								WAPE
Time-Steps	Days	#	Type	Blocks	Filters	Kernel	Dilations	Batch Size	Epochs	
168	7	7	LSTM	2	128			128	50	0.4317
		8	TCN	1	128	3	[1, 5, 7, 14, 28, 56]	128	50	0.4272
336	14	9	LSTM	3	64			64	50	0.4429
		10	TCN	2	128	3	[1, 5, 7, 14, 28, 56]	64	100	**0.4228**
672	28	11	LSTM	2	64			64	50	0.4336
		12	TCN	2	64	6	[1, 5, 7, 14, 28, 56]	64	50	0.4319

For the electric vehicle power consumption data, the best result (0.4228 WAPE) has been obtained with a past history of 14 days (336 time-steps). This TCN architecture, which is represented in Figure 7b, has two residual blocks with six convolutional layers of kernel size 3, 128 filters, and dilations of 1, 5, 7,

14, 28, and 56. This model has been trained over 100 epochs with a batch size of 64 instances. Similar to the previous dataset, the best LSTM model (0.04317 WAPE) uses the smallest possible input window, which is 7 days. This LSTM model consists of 2 stacked layers with 128 units and was trained over 50 epochs using 128 as batch size. Given the different nature of this dataset, the best configurations of architectures present several differences. For TCNs, a smaller kernel was able to extract the underlying patterns more accurately. Furthermore, a smaller batch size was better in almost all cases since the EV time series has fewer instances to train with.

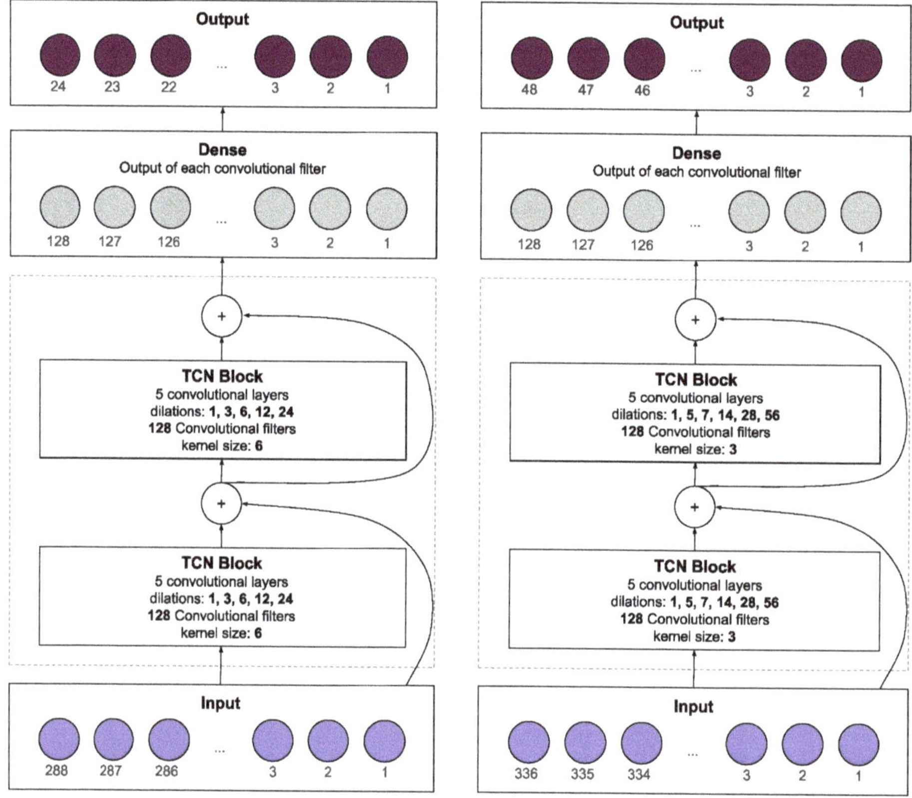

(**a**) Best TCN architecture for the electric demand data. (**b**) Best TCN architecture for the EV data.

Figure 7. Architecture of the best TCN model for each dataset.

In addition, Figure 8 shows the evolution of the training and validation loss for the best TCN and LSTM models. These plots can help to compare the learning process of the different architectures presented in Table 3. It can be seen that TCNs have a more stable loss optimisation procedure compared to the LSTM models, which can be associated with the use of the standard backpropagation method. The training curves of convolutional networks suffer fewer oscillations than the recurrent approach. For the electric demand dataset, TCN models converge faster to a lower validation loss. However, for the EV power demand data, LSTMs have an inferior validation loss at the initialisation point and hence converge more rapidly.

In general, the obtained results demonstrate that TCN models can outperform the forecasting accuracy of LSTM models. TCNs have shown a more reliable performance regardless of the selected architecture and parametrisation. The dilated casual convolutions employed by TCNs were better at capturing long-term dependencies than recurrent units. Other important conclusions can be drawn from the experimental results of this study. As can be seen, the best LSTM models for both datasets use the smallest possible input window. This indicates that the sequential processing of recurrent networks is not optimal for dealing with very long input sequences. In contrast, the use of residual connections in TCNs allows to increase the depth of the network and effectively encode longer sequences. Furthermore, another aspect to consider is that the best performance was obtained when stacked layers used as many filters or units as possible.

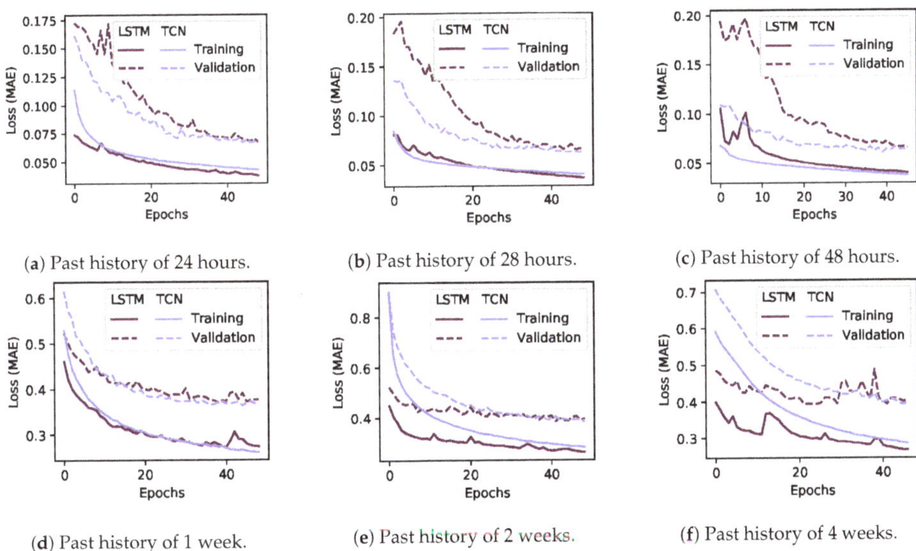

(**a**) Past history of 24 hours. (**b**) Past history of 28 hours. (**c**) Past history of 48 hours.

(**d**) Past history of 1 week. (**e**) Past history of 2 weeks. (**f**) Past history of 4 weeks.

Figure 8. Evolution of the training and validation loss function of the best TCN and LSTM models for each possible value of past history. First row corresponds to the electric demand data (**a–c**), while the second row corresponds to the EV power consumption data (**d–f**).

3.2. Computation Time

The deep learning models proposed in this study have complex architectures that involve a high computational cost. Therefore, it is also essential to evaluate the models in terms of computational efficiency. Table 4 presents the training and prediction time for all the best TCNs and LSTMs that were presented in the previous subsection. It also reports the number of trainable parameters to further enhance the comparative study. For obtaining these computation times, the batch size has been fixed to 128 instances for all models in order to perform fair comparisons. As can be seen, TCNs have far more trainable parameters since their stacked blocks have many convolutional operations. This results in higher training time compared to the LSTM models. The differences in training times are slightly larger in the case of the electric demand data, considering that an epoch of this dataset involves many more instances than the EV data. However, with respect to prediction time, TCN models are always faster than recurrent networks. Regardless of the considerably higher number of parameters, TCNs are able to provide very quick forecasts once they are trained. As it was expected, parallel convolutions of TCNs can process long input sequences faster than the sequential processing of recurrent networks. This seems an essential advantage for real-time applications, in which predictions need to be obtained as soon as possible in order to make informed decisions.

Table 4. Computation time of the best TCN and LSTM models for each dataset. Prediction time corresponds to the time that a model takes to compute a prediction for one instance. Training time corresponds to the time that a model takes to run one training epoch.

a Computation time for electric demand data.

Past History		Model		Trainable Paramereters	Prediction Time (ms)	Training Time (s)
Time-Steps	Hours	#	Type			
144	24	1	LSTM	332,824	0.859	**52.24**
		2	TCN	1,694,448	**0.566**	87.29
168	28	3	LSTM	332,824	0.968	**58.56**
		4	TCN	1,694,448	**0.590**	99.44
288	48	5	LSTM	201,240	0.775	**62.70**
		6	TCN	2,121,200	**0.672**	109.2

b Computation time for EV power consumption data.

Past History		Model		Trainable Paramereters	Prediction Time (ms)	Training Time (s)
Time-Steps	Days	#	Type			
168	7	7	LSTM	332,824	0.532	**3.01**
		8	TCN	1,694,448	**0.347**	4.41
336	14	9	LSTM	332,824	1.020	**6.76**
		10	TCN	1,694,448	**0.836**	10.49
672	28	11	LSTM	201,240	1.148	**4.76**
		12	TCN	2,121,200	**0.726**	10.18

4. Conclusions

In this paper, we proposed a deep learning model based on temporal convolutional networks (TCN) to perform forecasting over two energy-related time series. The experimental study considered two real-world time series data from Spain: the national electric demand and the power demand at charging stations for electric vehicles. An extensive parameter search was conducted in order to obtain the best architecture configuration, testing more than 1200 different TCN models for both dataset. Furthermore, the performance of these convolutional networks was compared in terms of accuracy and efficiency with long short-term memory (LSTM) recurrent networks—that have so far been considered the state-of-the-art for forecasting tasks.

The results of the experimental study carried out showed that TCNs outperformed the forecasting accuracy of LSTM models for both datasets. The dilated causal convolutions used by TCNs were more effective at capturing temporal dependencies than the recurrent LSTM units. Furthermore, TCNs proved to be less sensitive to the parameter selection than LSTM models. Regardless of the chosen values, the convolutional approach provided a more reliable performance. Moreover, we also aimed to illustrate the importance of the size of the past history input window. Thanks to the use of residual connections, TCNs provided better results when using longer input sequences. In contrast, LSTM models were more accurate at encoding patterns when using smaller windows.

Regarding the computational efficiency, it was seen that TCN models have deeper architectures with many more trainable parameters. This implied that the training procedure of a TCN was slightly more costly. However, once TCNs were trained, they provided significantly faster predictions than recurrent networks due to the use of parallel convolutions to process the input sequences. In conclusion, our study demonstrated that TCNs are a very powerful alternative to LSTM networks. They can provide more accurate predictions and are more suitable for real-time applications given their faster predicting speed.

Future efforts on this path will be focused on analysing the use of ensembles of TCN blocks with different receptive fields and using techniques such as evolutionary algorithms for the parameter search process. Another interesting future work could be the application of TCN networks in an online environment for real-time data streaming forecasting. Moreover, further research should also study the suitability of TCN networks for other problems like multivariable time series forecasting or time series classification.

Author Contributions: All authors made substantial contributions to conception and design of the study. P.L.-B. and M.C.-G. performed the experiments, analysed the data, and wrote the paper. J.M.L.-R. and J.C.R. guided the research and reviewed the manuscript. All authors have read and agreed to the published version of the manuscript.

Funding: This research has been funded by the Spanish Ministry of Economy and Competitiveness under the project TIN2017-88209-C2-2-R and by the Andalusian Regional Government under the projects: BIDASGRI: Big Data technologies for Smart Grids (US-1263341), Adaptive hybrid models to predict solar and wind renewable energy production (P18-RT-2778).

Acknowledgments: We are grateful to NVIDIA for their GPU Grant Program that has provided us the high-quality GPU devices for carrying out the study.

Conflicts of Interest: The authors declare no conflict of interest. The funders had no role in the design of the study; in the collection, analyses, or interpretation of data; in the writing of the manuscript, or in the decision to publish the results.

Abbreviations

The following abbreviations are used in this manuscript:

ANN	Artificial Neural Network
CNN	Convolutional Neural Network
DL	Deep Learning
EV	Electric vehicle
LSTM	Long Short-Term Memory Network
MAE	Mean Absolute Error
MIMO	Multi-Input Multi-Output
RNN	Recurrent Neural Network
SVM	Support Vector Machine
TCN	Temporal Convolutional Network
WAPE	Weighted Absolute Percentage Error

References

1. Contreras, J.; Espínola, R.; Nogales, F.; Conejo, A. ARIMA models to predict next-day electricity prices. *IEEE Trans. Power Syst.* **2003**, *18*, 1014–1020. [CrossRef]
2. Yang, Z.; Ce, L.; Lian, L. Electricity price forecasting by a hybrid model, combining wavelet transform, ARMA and kernel-based extreme learning machine methods. *Appl. Energy* **2017**, *190*, 291–305. [CrossRef]
3. Panapakidis, I.; Dagoumas, A. Day-ahead electricity price forecasting via the application of artificial neural network based models. *Appl. Energy* **2016**, *172*, 132–151. [CrossRef]
4. Kavousi-Fard, A.; Samet, H.; Marzbani, F. A new hybrid Modified Firefly Algorithm and Support Vector Regression model for accurate Short Term Load Forecasting. *Expert Syst. Appl.* **2014**, *41*, 6047–6056. [CrossRef]
5. Barman, M.; Dev Choudhury, N.; Sutradhar, S. A regional hybrid GOA-SVM model based on similar day approach for short-term load forecasting in Assam, India. *Energy* **2018**, *145*, 710–720. [CrossRef]

6. Fan, C.; Xiao, F.; Wang, S. Development of prediction models for next-day building energy consumption and peak power demand using data mining techniques. *Appl. Energy* **2014**, *127*, 1–10. [CrossRef]
7. Torres, J.; Troncoso, A.; Koprinska, I.; Wang, Z.; Martínez-Álvarez, F. Deep Learning for Big Data Time Series Forecasting Applied to Solar Power. In Proceedings of the 13th International Conference on Soft Computing Models in Industrial and Environmental Applications, San Sebastian, Spain, 6–8 June 2018; pp. 123–133. [CrossRef]
8. Ray, P.; Mishra, D.; Lenka, R. Short term load forecasting by artificial neural network. In Proceedings of the 2016 International Conference on Next Generation Intelligent Systems (ICNGIS), Kottayam, India, 1–3 September 2016. [CrossRef]
9. Schäfer, A.M.; Zimmermann, H.G. Recurrent Neural Networks Are Universal Approximators. In Proceedings of the 16th International Conference on Artificial Neural Networks ICANN'06, Athens, Greece, 10–14 September 2006; pp. 632–640. [CrossRef]
10. Wang, Y.; Liu, M.; Bao, Z.; Zhang, S. Short-Term Load Forecasting with Multi-Source Data Using Gated Recurrent Unit Neural Networks. *Energies* **2018**, *11*, 1138. [CrossRef]
11. Zheng, J.; Xu, C.; Zhang, Z.; Li, X. Electric load forecasting in smart grids using Long-Short-Term-Memory based Recurrent Neural Network. In Proceedings of the 2017 51st Annual Conference on Information Sciences and Systems (CISS), Baltimore, MD, USA, 22–24 March 2017; pp. 1–6. [CrossRef]
12. Hochreiter, S.; Schmidhuber, J. Long Short-term Memory. *Neural Comput.* **1997**, *9*, 1735–1780. [CrossRef]
13. Ma, X.; Tao, Z.; Wang, Y.; Yu, H.; Wang, Y. Long short-term memory neural network for traffic speed prediction using remote microwave sensor data. *Transp. Res. Part C Emerg. Technol.* **2015**, *54*, 187–197. [CrossRef]
14. Gensler, A.; Henze, J.; Sick, B.; Raabe, N. Deep Learning for solar power forecasting—An approach using AutoEncoder and LSTM Neural Networks. In Proceedings of the 2016 IEEE International Conference on Systems, Man, and Cybernetics (SMC), Budapest, Hungary, 9–12 October 2016; pp. 2858–2865. [CrossRef]
15. Fischer, T.; Krauss, C. Deep learning with long short-term memory networks for financial market predictions. *Eur. J. Oper. Res.* **2018**, *270*, 654–669. [CrossRef]
16. Peng, L.; Liu, S.; Liu, R.; Wang, L. Effective long short-term memory with differential evolution algorithm for electricity price prediction. *Energy* **2018**, *162*, 1301–1314. [CrossRef]
17. Amarasinghe, K.; Marino, D.; Manic, M. Deep neural networks for energy load forecasting. In Proceedings of the 2017 IEEE 26th International Symposium on Industrial Electronics (ISIE), Edinburgh, UK, 19–21 June 2017; pp. 1483–1488. [CrossRef]
18. Almalaq, A.; Edwards, G. A review of deep learning methods applied on load forecasting. In Proceedings of the 2017 16th IEEE International Conference on Machine Learning and Applications (ICMLA), Cancun, Mexico, 18–21 December 2017; pp. 511–516. [CrossRef]
19. Kuo, P.H.; Huang, C.J. A High Precision Artificial Neural Networks Model for Short-Term Energy Load Forecasting. *Energies* **2018**, *11*, 213. [CrossRef]
20. Koprinska, I.; Wu, D.; Wang, Z. Convolutional Neural Networks for Energy Time Series Forecasting. In Proceedings of the 2018 International Joint Conference on Neural Networks (IJCNN), Rio de Janeiro, Brazil, 8–13 July 2018; pp. 1–8. [CrossRef]
21. Cirstea, R.G.; Micu, D.V.; Muresan, G.M.; Guo, C.; Yang, B. Correlated Time Series Forecasting Using Multi-Task Deep Neural Networks. In *Proceedings of the 27th ACM International Conference on Information and Knowledge Management*; ACM: New York, NY, USA, 2018; pp. 1527–1530. [CrossRef]
22. Tian, C.; Ma, J.; Zhang, C.; Zhan, P. A Deep Neural Network Model for Short-Term Load Forecast Based on Long Short-Term Memory Network and Convolutional Neural Network. *Energies* **2018**, *11*, 3493. [CrossRef]
23. Shen, Z.; Zhang, Y.; Lu, J.; Xu, J.; Xiao, G. A novel time series forecasting model with deep learning. *Neurocomputing* **2019**. [CrossRef]
24. Bai, S.; Kolter, J.Z.; Koltun, V. An Empirical Evaluation of Generic Convolutional and Recurrent Networks for Sequence Modeling. *arXiv* **2018**, arXiv:1803.01271.
25. Borovykh, A.; Bohte, S.; Oosterlee, C. Dilated convolutional neural networks for time series forecasting. *J. Comput. Financ.* **2019**, *22*, 73–101. [CrossRef]
26. Chen, Y.; Kang, Y.; Chen, Y.; Wang, Z. Probabilistic Forecasting with Temporal Convolutional Neural Network. *Neurocomputing* **2020**. [CrossRef]

27. Wan, R.; Mei, S.; Wang, J.; Liu, M.; Yang, F. Multivariate temporal convolutional network: A deep neural networks approach for multivariate time series forecasting. *Electronics* **2019**, *8*, 876. [CrossRef]
28. Spanish Public Grid (Red eléctrica de España). Available online: https://www.ree.es/es/datos/ (accessed on 25 February 2020).
29. Galicia, A.; Torres, J.F.; Martínez-Álvarez, F.; Troncoso, A. Scalable Forecasting Techniques Applied to Big Electricity Time Series. In *Advances in Computational Intelligence*; Rojas, I., Joya, G., Catala, A., Eds.; Springer International Publishing: Cham, Switzerland, 2017; pp. 165–175.
30. Cecovel: Centro de control de vehículo eléctrico (Monitoring Centre of Electric Vehicles). Available online: https://www.ree.es/es/red21/vehiculo-electrico/cecovel (accessed on 25 February 2020).
31. Gómez-Quiles, C.; Asencio-Cortés, G.; Gastalver-Rubio, A.; Martínez-Álvarez, F.; Troncoso, A.; Manresa, J.; Riquelme, J.C.; Riquelme-Santos, J.M. A Novel Ensemble Method for Electric Vehicle Power Consumption Forecasting: Application to the Spanish System. *IEEE Access* **2019**, *7*, 120840–120856. [CrossRef]
32. Taieb, S.B.; Bontempi, G.; Atiya, A.F.; Sorjamaa, A. A review and comparison of strategies for multi-step ahead time series forecasting based on the NN5 forecasting competition. *Expert Syst. Appl.* **2012**, *39*, 7067–7083. [CrossRef]
33. Bandara, K.; Bergmeir, C.; Smyl, S. Forecasting across time series databases using recurrent neural networks on groups of similar series: A clustering approach. *Expert Syst. Appl.* **2020**, *140*, 112896. [CrossRef]
34. Petersen, N.C.; Rodrigues, F.; Pereira, F.C. Multi-output bus travel time prediction with convolutional LSTM neural network. *Expert Syst. Appl.* **2019**, *120*, 426–435. [CrossRef]
35. Yu, F.; Koltun, V. Multi-Scale Context Aggregation by Dilated Convolutions. In Proceedings of the 4th International Conference on Learning Representations, ICLR 2016, San Juan, Puerto Rico, 2–4 May 2016.
36. Lara-Benítez, P.; Carranza-García, M.; García-Gutiérrez, J.; Riquelme, J. Asynchronous dual-pipeline deep learning framework for online data stream classification. *Integr. Comput.-Aided Eng.* **2020**, 1–19. [CrossRef]
37. Nair, V.; Hinton, G. Rectified linear units improve Restricted Boltzmann machines. In Proceedings of the ICML 2010—27th International Conference on Machine Learning, Haifa, Israel, 21–24 June 2010; pp. 807–814.
38. Van den Oord, A.; Dieleman, S.; Zen, H.; Simonyan, K.; Vinyals, O.; Graves, A.; Kalchbrenner, N.; Senior, A.W.; Kavukcuoglu, K. WaveNet: A Generative Model for Raw Audio. *arXiv* **2016**, arXiv:1609.03499.
39. He, K.; Zhang, X.; Ren, S.; Sun, J. Deep Residual Learning for Image Recognition. In Proceedings of the IEEE Conference on Computer Vision and Pattern Recognition, Las Vegas, NV, USA, 27–30 June 2016; pp. 770–778. [CrossRef]
40. Pascanu, R.; Mikolov, T.; Bengio, Y. On the difficulty of training Recurrent Neural Networks. *arXiv* **2012**, arXiv:cs.LG/1211.5063.
41. Kingma, D.P.; Ba, J. Adam: A Method for Stochastic Optimization. *arXiv* **2014**, arXiv:cs.LG/1412.6980.
42. Lara-Benítez, P.; Carranza-García, M. Electric Demand Forecasting with Temporal Convolutional Networks. Available online: https://github.com/pedrolarben/ElectricDemandForecasting-DL (accessed on 1 March 2020).
43. Keskar, N.S.; Mudigere, D.; Nocedal, J.; Smelyanskiy, M.; Tang, P.T.P. On Large-Batch Training for Deep Learning: Generalization Gap and Sharp Minima. *arXiv* **2016**, arXiv:1609.04836.

MDPI
St. Alban-Anlage 66
4052 Basel
Switzerland
Tel. +41 61 683 77 34
Fax +41 61 302 89 18
www.mdpi.com

Applied Sciences Editorial Office
E-mail: applsci@mdpi.com
www.mdpi.com/journal/applsci

www.ingramcontent.com/pod-product-compliance
Lightning Source LLC
LaVergne TN
LVHW070615170726
843515LV00004B/88

* 9 7 8 3 0 3 6 5 0 8 6 2 7 *